VIRTUAL COMPONENTS DESIGN AND REUSE

Virtual Components Design and Reuse

Edited by

Ralf Seepold

and

Natividad Martínez Madrid
Forschungszentrum Informatik Karlsruhe (FZI), Germany

KLUWER ACADEMIC PUBLISHERS
BOSTON / DORDRECHT / LONDON

A C.I.P. Catalogue record for this book is available from the Library of Congress.

ISBN 0-7923-7261-1

Published by Kluwer Academic Publishers,
P.O. Box 17, 3300 AA Dordrecht, The Netherlands.

Sold and distributed in North, Central and South America
by Kluwer Academic Publishers,
101 Philip Drive, Norwell, MA 02061, U.S.A.

In all other countries, sold and distributed
by Kluwer Academic Publishers,
P.O. Box 322, 3300 AH Dordrecht, The Netherlands.

Printed on acid-free paper

All Rights Reserved
© 2001 Kluwer Academic Publishers, Boston
No part of the material protected by this copyright notice may be reproduced or
utilized in any form or by any means, electronic or mechanical,
including photocopying, recording or by any information storage and
retrieval system, without written permission from the copyright owner.

Printed in the Netherlands.

TABLE OF CONTENTS

LIST OF FIGURES		IX
LIST OF TABLES		XIII

1 VIRTUAL COMPONENTS - FROM RESEARCH TO BUSINESS 1
Jürgen Haase
  1.1 Application of Virtual Components 2
  1.2 The Virtual Components Business 7
  1.3 Research in the Reuse Field 10
  1.4 Conclusions 11

2 EVALUATION OF TECHNOLOGY AND THE MEDEA DESIGN AUTOMATION ROADMAP 13
Joseph Borel, Anton Sauer and Ralf Seepold
  2.1 Introduction 13
  2.2 Targets of EDA Activities 14
  2.3 The Semiconductor Industry Challenge 16
  2.4 Roadmap Coverage 19
  2.5 IP Reuse Roadmap 22
  2.6 Conclusion 32

3 PRODUCTIVITY IN VC REUSE: LINKING SOC PLATFORMS TO ABSTRACT SYSTEMS DESIGN METHODOLOGY 33
Grant Martin
  3.1 Overview 33
  3.2 The Nature of System Design 34
  3.3 SOC Integration Platforms 34
  3.4 Function-Architecture Co-Design 36
  3.5 Linking the two Concepts: Platform-Based System Co-Design 39
  3.6 Multimedia Examples 43
  3.7 Other Applications of Platform-Based System Design 45
  3.8 Conclusion and Future 46

4 SOFTWARE IP IN EMBEDDED SYSTEMS 47
Carsten Böke, Carsten Ditze, H.J. Eickerling, Uwe Glässer, Bernd Kleinjohann, Franz Rammig and Wolfgang Thronicke
  4.1 Introduction 48
  4.2 CORBA: The Base Technology for Integration 49
  4.3 Types of Coupling for Software IP 51
  4.4 Bridging Semantics: The ASM Approach 53

	4.5 Reuse and Workflows - towards Intelligent Resource Management	55
	4.6 Document Exchange in Software IP Frameworks	56
	4.7 An Application Example: Building Real-Time Communication Systems from Software IP	58

5 ARDID: A TOOL AND A MODEL FOR THE QUALITY ANALYSIS OF VHDL BASED DESIGNS 65

Yago Torroja, Felipe Machado, Fernando Casado, Eduardo de la Torre, Teresa Riesgo and Javier Uceda

5.1	Quality Attributes of VHDL Designs	65
5.2	ARDID: A VHDL Quality Analysis Tool	67
5.3	Inside ARDID: Simplified Hardware Model	73
5.4	Experimental Results	77
5.5	Conclusions	79

6 A VHDL ANALYSIS ENVIRONMENT FOR DESIGN REUSE 81

Claudio Costi and D. Michael Miller

6.1	Introduction	81
6.2	Reuse of Soft Components	82
6.3	Valet Project	83
6.4	From VHDL Source Code to Design Knowledge	85
6.5	Conclusions	94

7 LAMBDA-BLOCK ANALYSIS OF VHDL FOR DESIGN REUSE 95

William Fornaciari, Salvatore Minonne, Fabio Salice and Massimo Vincenzi

7.1	Introduction	95
7.2	Overview of the global Environment	97
7.3	Testbench Orthogonality	98
7.4	Lambda-blocks Analysis	99
7.5	The Reuse Experiment	102
7.6	Concluding Remarks	103

8 IP RETRIEVAL BY SOLVING CONSTRAINT SATISFACTION PROBLEMS 105

Manfred Koegst, Jörg Schneider, Ralph Bergmann and Ivo Vollrath

8.1	Introduction	105
8.2	Archiving and Retrieval in the Reuse Process	106
8.3	Modelling and Retrieval by case-based Reasoning	108
8.4	Concept of an Adapted Database Management	112
8.5	Summary	115
8.6	Example	116

9 CRYPTOGRAPHIC REUSE LIBRARY 119
Andreas Schubert, Ralf Jährig and Walter Anheier
- 9.1 Introduction 119
- 9.2 Library Concept 120
- 9.3 CVC Structure 123
- 9.4 CVC interfaces 127
- 9.5 Conclusions 129

10 A VHDL REUSE COMPONENT MODEL FOR MIXED ABSTRACTION LEVEL SIMULATION AND BEHAVIORAL SYNTHESIS 131
Cordula Hansen, Oliver Bringmann and Wolfgang Rosenstiel
- 10.1 Introduction 131
- 10.2 A VHDL Reuse Component Model for Simulation and Synthesis 132
- 10.3 The Frame Component Concept 134
- 10.4 Reuse Component Library and CADDY-II 136
- 10.5 Frame Component and Synthesis 138
- 10.6 Frame Component and Simulation 140
- 10.7 Conclusion 144

11 VIRTUAL COMPONENT INTERFACES 145
M. M. Kamal Hashmi
- 11.1 Introduction 145
- 11.2 Definitions and Terminology 146
- 11.3 VC Creation and ReUse 146
- 11.4 Cutting the Cost of ReUse 147
- 11.5 Interface Based Design 148
- 11.6 VCI Specification Standards Efforts 154
- 11.7 VSI SLD Interface Documentation Standard 154
- 11.8 VHDL Extensions for Interfaces and System Design 156
- 11.9 Conclusion 157
- 11.10 Acknowledgements 157

12 A METHOD FOR INTERFACE CUSTOMIZATION OF SOFT IP CORES 159
Robert Siegmund and Dietmar Müller
- 12.1 Introduction 159
- 12.2 Method 161
- 12.3 Interface Synthesis 166
- 12.4 Results 169
- 12.5 Conclusions and Future Work 170

13 MODELING ASSISTANT - A FLEXIBLE VCM GENERATOR IN VHDL 171
Andrzej Pulka
13.1 Introduction 171
13.2 Reusability Of VHDL Models 171
13.3 Data Model 172
13.4 Generation Entity of Model (GEM) 176
13.5 Experiments and Results 177
13.6 Summary 180

14 REUSING IPS TO IMPLEMENT A SPARC® SOC 183
Serafín Olcoz, Alfredo Gutiérrez and Denis Navarro
14.1 Introduction 183
14.2 SPARC μProcessor IP 184
14.3 μProcessor Peripherals for Embedded Systems 188
14.4 A μController for a Scalable Remote Terminal Unit 189
14.5 CoMES: Co-Design Methodology for Embedded Systems 191
14.6 Conclusion and Future Work 192

15 HARDWWWIRED: USING THE WEB AS REPOSITORY OF VHDL COMPONENTS 195
Adriano Sarmento, Jorge Fernandes and Edna Barros
15.1 Introduction 195
15.2 HARDWWWIRED: The Architecture 197
15.3 Mapping VHDL Descriptions Into JAVA Objects 198
15.4 Security Policies 199
15.5 Designing a CPU With HARDWWWIRED: A Case Study 201
15.6 Conclusion and Future Work 204

REFERENCES 207

INDEX 223

GLOSSARY 227

LIST OF FIGURES

Figure 1.1	Phases of the reuse history	2
Figure 1.2	Set-top-box with SICAN DesignObjectsTM	3
Figure 1.3	SICAN's House of Methodology	5
Figure 1.4	Application of formal verification	6
Figure 1.5	Designs with reusable IP	7
Figure 1.6	Percentage of ASIC market with and without IP content	7
Figure 1.7	The importance of reuse and IP business	9
Figure 2.1	Time to market revenue penalties	15
Figure 2.2	Technology platform	16
Figure 2.3	Process design gap	17
Figure 2.4	Innovation and maintenance cycle	18
Figure 2.5	Classical IC products design	19
Figure 2.6	Analogue/digital design and test flow	20
Figure 2.7	Roadmap forecast	21
Figure 2.8	Roadmap for IP reuse	25
Figure 3.1	Functional model of voice mail pager	36
Figure 3.2	Possible architecture for pager application	37
Figure 3.3	Mapping of voice pager behavior to architecture	38
Figure 3.4	JPEG encoder-decoder mapped onto the platform	41
Figure 3.5	JPEG decoder and voice mail pager derivative application	42
Figure 3.6	Results of two different mappings and prioritizations	43
Figure 3.7	Basic multimedia SOC integration platform	44
Figure 3.8	VOP application for platform	44
Figure 3.9	Audio on demand application mapped to the platform	45
Figure 4.1	Language coupling	51
Figure 4.2	"Depot" skeleton for the producer/consumer problem	60
Figure 4.3	Synthesis process for component-based software	63
Figure 5.1	Outlook of the ARDID environment	72
Figure 5.2	Simplified hardware model	73
Figure 5.3	Differences between synthesis and SHM for Table 5.4	77
Figure 6.1	VALET tool environment	85
Figure 6.2	FPDG graphs of a design	88
Figure 7.1	The overall environment to analyze design reusability	97
Figure 7.2	The lambda-block analysis flow	100
Figure 7.3	Forward activation (left) and backward propagation (right)	101
Figure 7.4	Functionality extraction via intersection of l-blocks	101
Figure 8.1	Scheme for IP archiving and retrieval	108
Figure 8.2	Scheme for deciding a generalized CSDP (Q,c)	111

Figure 8.3	Three-layer-model (left) and structure of the Server-Host (right)	113
Figure 8.4	Entity-Relationship-Model	114
Figure 8.5	Part of the structure of an 8051-IP (FhG-IIS-Erlangen)	115
Figure 9.1	Cryptographic virtual component (CVC)	120
Figure 9.2	Increasing application flexibility by optional wrapper modules	120
Figure 9.3	Procedure of CVC hardware configuration	121
Figure 9.4	Simplified VLSI architecture of the algorithm core	123
Figure 9.5	Hardware configuration of the module (left)	126
Figure 9.6	Hardware configuration of the VI wrapper (right)	126
Figure 9.7	Internal interfaces	127
Figure 9.8	Procedure for connection of FSMs at different hierarchy levels	128
Figure 9.9	(a) Virtual interface of the CVC (b) Reusable I/O module	128
Figure 10.1	Hierarchical specification using access procedures	133
Figure 10.2	Interface relations	134
Figure 10.3	Frame component for a simple GCD example	136
Figure 10.4	Synthesis flow using CADDY-II and the VHDL preprocessor	137
Figure 10.5	Allocating user defined components	138
Figure 10.6	Simulation flow using the VHDL preprocessor	140
Figure 11.1	Hierarchic partitioning with interfaces	149
Figure 11.2	Static interface + dynamic interface	150
Figure 11.3	Working at multiple levels of abstraction	151
Figure 11.4	Bridging the Gap between the Function and the Architecture	153
Figure 11.5	The Virtual Component Interface	155
Figure 11.6	Interface level translation – example	157
Figure 12.1	IP selection and instantiation process in IP based system design	160
Figure 12.2	Principle of IP interface customization	161
Figure 12.3	Interface customization for Soft- and Hard IP without TLI	166
Figure 12.4	DFA and ICM state transition graph for transaction 'io_write'	168
Figure 12.5	IP area and performance for different interfaces	170
Figure 12.6	Customization Effort for different IP Interfaces	170
Figure 13.1	Modeling Assistant - the system of modeling for reusability	173
Figure 13.2	Examples of menu system of the Modeling Assistant	174
Figure 13.3	Main menu of the NVMG generator	175
Figure 13.4	Example of component specialization	175
Figure 13.5	Testbenches reuse and simulation results of Gray coder	178
Figure 13.6	Simulation results of track-hold model	180
Figure 13.7	Example of 4th-row low-pass filter with 2 reused blocks	180
Figure 13.8	Simulation results of track-hold model	181
Figure 14.1	Decreasing percentage of new design content in a SOC	184
Figure 14.2	IDeAS project CAD Flow	186
Figure 14.3	(A) HW-SW Co-simulation. (B) Regression test method used	187

Figure 14.4	Block diagram of ECU SOC.		190
Figure 14.5	ECU SOC design process		190
Figure 14.6	EDS architecture		191
Figure 14.7	VHDL-ICE architecture		192
Figure 14.8	EDS / VHDL-ICE communication		193
Figure 15.1	HardWWWired Architecture		197
Figure 15.2	Entity BNF mapping to Java objects		200
Figure 15.3	Simplified CPU architecture		201
Figure 15.4	Creating a new component		203
Figure 15.5	Schematic edition with automatic VHDL code generation		204
Figure 15.6	Simulation input and output		205

LIST OF TABLES

Table 1.1	Differing approaches to reuse projects	6
Table 5.1	Different usage for the elements of an array	74
Table 5.2	Loop statement	75
Table 5.3	Generate statements	76
Table 5.4	If statement covering all options without else	77
Table 5.5	Results from QTK for some designs	78
Table 5.6	Results from VTK for some designs	79
Table 6.1	A VHDL design example	86
Table 6.2	Signal analysis results	91
Table 6.3	Functional analysis results	93
Table 7.1	Analysis of the VHDL concurrent statements	102
Table 7.2	Lambda-blocks extracted and grouped by functionality	103
Table 7.3	Manpower distribution	103
Table 9.1	Typical hardware and software configuration areas for CVCs	121
Table 9.2	Algorithm properties of the used symmetric block ciphers	122
Table 9.3	Architecture features of the algorithm soft cores	123
Table 9.4	Performance and cost features of the firm cores (0.7 µm CMOS)	124
Table 9.5	Software configuration of the module of modes of operation	125
Table 9.6	Software configuration of the VI wrapper	126
Table 10.1	Frame component for a simple GCD example	135
Table 10.2	Reuse component library containing frame components	137
Table 10.3	Frame component with synthesis attributes	139
Table 10.4	Specification using an access procedure	141
Table 10.5	VHDL specification after preprocessing	143
Table 12.1	IP interface specification at transaction level	163
Table 12.2	Specification of a 2-wire serial protocol for transaction	164
Table 12.3	Behavioural specification using a TLI and interface transactions	165
Table 12.4	Mapping of VHDL+ interface statements to NDFA	167
Table 12.5	VHDL template for replacement	169
Table 13.1	Fragment of the library contents - examples of listframes.	176
Table 13.2	Examples of the GEM syntax.	177
Table 13.3	Fragment of the VITAL Level 0 entity of the AMD2901	177
Table 13.4	Fragment of entity description of the track-hold model.	179
Table 13.5	Examples of generated library of VITAL templates	182
Table 15.1	CPU Instruction Set	202

1 VIRTUAL COMPONENTS - FROM RESEARCH TO BUSINESS

Jürgen Haase

Sican GmbH
Hannover, Germany

Abstract

The application of Virtual Components (IP modules) and reuse has already been accepted as a prerequisite for designing Systems-on-Chip in the near future or even today. Many topics related to reuse and Virtual Components are no longer of pure academic interest, they now have impact on product developments and business. Beyond that, Virtual Components are becoming a business of its own. The application of reuse and Virtual Components in real product designs together with the progress of technology leads to new challenges for research in that field.

This chapter gives an overview about the research activities as well as commercial applications related to Virtual Components. Starting with a survey of *design-ins*[1] of Virtual Components in actual product developments the chapter will describe the application in different types of companies, the approach for applying Virtual Components in different countries and the key characteristics of today's reuse projects.

Based on market research in the ASIC field the business related to Virtual Components will be analyzed. The discussion of research activities in the reuse field covers research work already applied in real life projects, hot topics in research on Virtual Components and reuse and what are the challenges for research in order the push reuse of IP. Examples include different levels of formal verification, behavioral descriptions, IP taxonomy etc.

---

1. Integration of IP module in ASIC design.

1.1 APPLICATION OF VIRTUAL COMPONENTS

Figure 1.1 shows four phases of the reuse history. In the beginning, the design community hoped that reuse is the solution for the severe design gap problem. First research results and practical experience provided key enabling methods for reuse ([Haas99b], [Seep99a], [Mous99], [Schl99], [Haas99a], [Eise99], [Meye99], [Bran99]). One essential conclusion was that it requires 2.5 to 3 times the effort of a conventional design to develop a reusable Virtual Component (IP). First companies started to become IP vendors, 1st generation IP was produced. After the first enthusiasm and hype also disillusion came up, at the moment we have the transition to 2nd generation IP. See Section 1.2.3 for a closer look on phases of the reuse history.

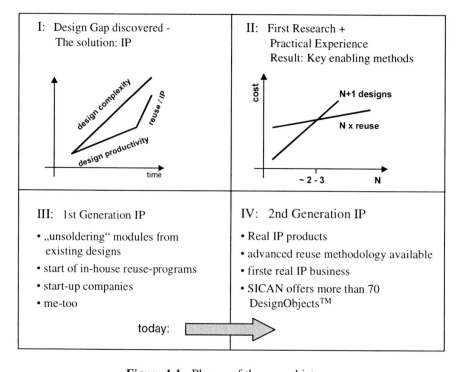

Figure 1.1 Phases of the reuse history

1.1.1 Example: Application Set-Top-Box

SICAN very successfully offers more than 70 DesignObjects™ (their Virtual Components) [Sica99a] with set-top-boxes belonging to the most popular applications with numerous design-ins. Thus set-top-boxes can serve as a well suited example for the application of Virtual Components.

VIRTUAL COMPONENTS - FROM RESEARCH TO BUSINESS 3

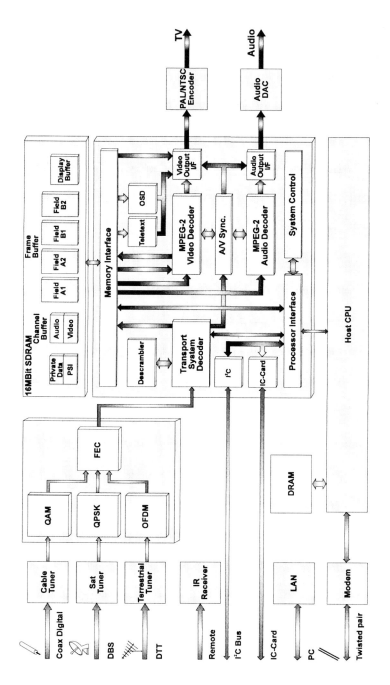

Figure 1.2 Set-top-box with SICAN DesignObjects™

4   VIRTUAL COMPONENTS DESIGN AND REUSE

Figure 1.2 shows a block diagram of a set-top-box with 25 functions required, 15 of which are productized as SICAN DesingObjects™ [Sica99a]. The following examples of a set-top-box development can serve as placeholders for typical classes of reuse projects:

Project example 1: IP user does the chip design himself
- IP user has most of the modules available or develops them himself
- IP user has enough engineering resources
- Only some modules (not belonging to the core know-how) will be purchased, e.g. for interfacing
- As example: I^2C, IC-Card interfaces will be purchased from the IP vendor

Project example 2: IP user outsources the chip design (completely or partly)
- IP user has not the required know-how or
- IP user has no appropriate resources available
- time-to-market is crucial
- Minimizing risk
- Minimum number of IP-vendors for the design desired

=> IP vendor should offer a complete solution
=> IP vendor must be able to provide extensive design support
=> Very difficult for small IP providers

Project example 3: IP user requires a complete solution for his product
- Whole chip design as described in example 2
- Plus software development (SW for embedded cores, drivers, APIs, ...)
- Plus system development: Evaluation boards, reference designs, ...

=> Can only be provided by large IP providers / design houses

1.1.2 Design Methodology

In order to be able to produce top-quality IP-products and to provide sophisticated design services to the customers, SICAN has developed its "House of Methodology" shown in Figure 1.3.

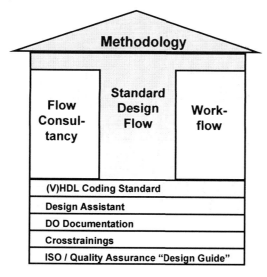

Figure 1.3 SICAN's House of Methodology

The foundation of the house are standards for IP products (HDL Coding Standards, DesignObjects^{TM} Documentation Standards [Sica99a], Quality Assurance) as well as means providing the individual designer's knowledge to all designers (Crosstrainings and the Design Assistant, a kind of cookbook). Based on this foundation there are three pillars of successful designs and IP products: a standard design flow mandatory for all designs, a workflow for production of DesignObjects^{TM} [Haas99b] and the flow consultancy team. The flow consultancy team supports each project from the beginning (starting with project set-up, installation of database) and provides leading edge methodology to all internal projects (turnkey designs, IP product developments), like a consulting service also to all customers.

E.g. formal verification was introduced as new methodology by SICAN experts. The flow consultancy team, supported by the cross-training, made this knowledge available to all SICAN designers. The first application was a set-top-box chip design, the methodology shown in Figure 1.4 resulted in an impressive performance:

- Design-in of DesignObjects^{TM} with several hundreds of Kgates
- Automated translation from VHDL to VerilogHDL
- Delivery less than 12 weeks after purchase order
- No bugs reported

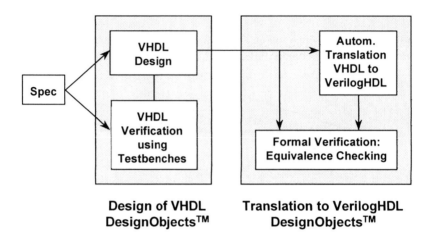

Figure 1.4 Application of formal verification

1.1.3 Approach to purchase Virtual Components in different countries

Table 1.1 summarizes SICAN's experience in reuse projects world-wide.

Europe	USA	Japan
• Looking for partnerships for complete solutions • Purchase of only few IP modules • Still very sceptical, paradigm shift has not really taken place yet • IP modules have to match exactly, not willing to design mainly with the IP available on the market	• Made experience using IP, also with buggy IP -> focus: minimizing risk • Clear commitment to using external IP • Virtual Components are looked upon as normal products -> extensive product documentation is expected to be available immediately • Willingness to design around available IP	• It takes a very long time to build up trust • Due to different cultures reuse business is difficult at the moment • Acquisition phase is long-lasting and requires very extensive communication and discussions

Table 1.1 Differing approaches to reuse projects

1.2 THE VIRTUAL COMPONENTS BUSINESS

1.2.1 What do the analysts tell?

An indicator for the business related to Virtual Components is the number of designs using Virtual Components. Figure 1.5 gives a forecast of the designs with reusable IP. This number is growing by more than 40% per year.

According to the growing number of designs also the market share of IP-based designs is growing rapidly. Figure 1.6 (Source: ICE ASIC Report 1999) shows how much of the overall ASIC market includes IP content. The trend in 1999 is that most larger ASIC designs include IP and that the IP business starts to take off. Future Horizon predicts that design houses doing high-level-chip-design (fabless) will grow by 34% per year until 2004.

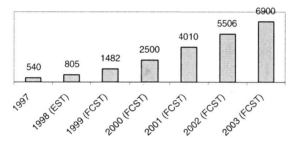

Figure 1.5 Designs with reusable IP

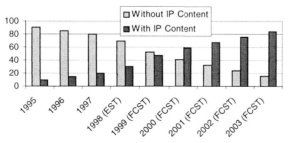

Figure 1.6 Percentage of ASIC market with and without IP content

1.2.2 How far is business in reality?

The market research results given in Section 1.2.1 promise a very prosperous future for the IP business. But how far is the IP business in reality today? First disillusions about the IP business can be observed as well as first success stories. E.g. SICAN as Europe's largest independent design house and one of the pioneers in the IP business has passed in 1999 the break even point in its DesignObjects^{TM} (Virtual Components) business.

The actual situation in the main types of IP-related business is as follows:

Intra-company reuse:
- All major companies have corporate reuse programs in place, often already including internal business models
- The concepts are (more or less) ready for application
- The paradigm shift associated with the concept of reuse still has not taken place in the design departments, most designers don't practice design-by-reuse at the moment (the required change of mind set is a difficult task)

=> Intra-company reuse is not a real business yet

IP-Providers: Semiconductor companies and EDA companies:
- Selling IP supports the core business (selling silicon/tools)

=> IP-products do have a relevant impact on business for these companies, even if they are not a real business of its own

IP-Providers: Brokers of IP:
- The quality of the offered IP varies a lot, design support often is not available, the added value offered by IP brokers at the moment is not sufficiently high

=> Brokering IP is no real business yet (but might be soon)

IP-Providers: Start-ups, small IP-selling companies:
- Some are successful
- Many had to give up

=> successful ones will be taken over (often that's their vision)

IP-Providers: big independent IP vendors (e.g. SICAN):
- A substantial amount of design services must be available
- Complete system solutions required (seeSection 1.1)
- Customers expect real products including support and maintenance
- Many customers prefer big players in order to reduce risk

=> IP vendors fulfilling these prerequisites have already been able to pass the break-even point

1.2.3 The importance of IP business: hype - disillusion - hope

The importance of IP business changed over the years, not always the prevailing opinion and the reality are matched, as illustrated in Figure 1.7.

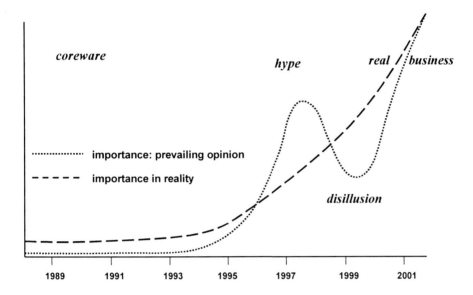

Figure 1.7 The importance of reuse and IP business

The genesis of reuse and IP business can be broken down into four phases:

Phase 1:

About 10 years ago the fields reuse and IP were no real issues in the design community, the significance of them was judged to be very low. However, in reality reuse already began to take place as semiconductor companies offered some (what we today call) Virtual Components, e.g. they called it coreware. Thus, in reality the importance of reuse and IP was higher than expected.

Phase 2: Hype and me-too

Starting about 1995 reuse became a very popular topic, in the academic world as well as in industry. Many people became very enthusiastic, they thought major problems like the design gap to be solved by reuse, and finally a lot of people and companies saw a lot of Euros and Dollars within reach for them by selling IP. Thus many said "me-too", started stating "we provide IP", "we are doing reuse" etc. Actually, many of them were just doing the same they did before, their IP were just modules "unsolded" from an existing chip design. In reality, practical applications of IP and reuse remained behind this enthusiasm, but made some progress.

Phase 3: Disillusion

As a result of phase 2 many people became very disillusioned about IP, as they made their first experience with the problems of making reuse a reality, by purchasing buggy IP etc. The estimations about the importance of IP and the related business opportunities declined, revenues with IP were not as high as expected, a number of start-up companies had to give up, some people even declared the IP business not to exist any more.

However, at the same time the real IP business became reality for some companies, who had the right concepts and who did the hard work which is essential for being an IP vendor. Examples are ARM and SICAN, both of them selling very successful IP products as well as design services. SICAN introduced their DesignObjects^{TM} (Virtual Components) in 1996 and has passed the break even point in this business in 1999.

Phase 4: Hope

In the near future the IP vendors as well as their customers will do their homework and will make substantial progress in filling the gap in their design methodology. More and more standard design flows will include design-in of IP modules, IP definitely will become a business of its own.

1.3 RESEARCH IN THE REUSE FIELD

As research on Virtual Components and reuse is a rather young discipline, no overall theory is available. Encouraging progress has been made in some aspects, but in many other parts it has been more or less "trial and error". In many cases "reuse" was just a new label for already existing work without developing really new approaches.

True reuse-research has just begun, a lot of work remains to be done. At the moment research is working on the required methods and tools for the next generation of reuse and IP products, enabling real IP business. Many of the hot topics in reuse-research are covered in this book, important fields are:

- High level design methodology

  Examples are methods allowing for taking key design decisions already at system level and not at an implementation level [Mart99b], adaptive interfaces of soft IP cores [Sieg99] and the application of C/C++ language for system design including synthesis etc.

- Prototyping of SoC designs

  For the very difficult task of verifying SoC designs prototyping is important. The approaches include virtual prototyping for HW/SW co-design (HW/SW co-design will become very important for reuse in general) and programma-

ble solutions covering all phases of design (virtual prototyping, HW prototypes, product).
- Analyzing designs and design quality

   Topics are means and tools for analyzing design quality [Torr99], metrics for the degree of reusability [Bern99] and analysis of designs by applying SW engineering methods [Cost99].
- IP repositories and retrieval of IP, classification, metrics [Seep99a], [Koeg99]
- IP protection, watermarking
- Qualification of IP

   One of the most severe problems for IP users as well as for IP providers. Very promising seems to be the formal verification approach, e.g. model checking might be used for replacing functional evaluation of an IP.

In addition to these reuse-specific topics the progress of semiconductor technology requires severe changes in design methodology, as effects like crosstalk, EMC etc. will become very significant. Today's approach of totally separated phases of the design (HDL coding, synthesis, layout etc.) will not be appropriate any more. E.g. the layout will lead to changes on architectural level.

1.4 CONCLUSIONS

The chapter gave an overview about the actual status of research and applications in the Virtual Components field. The forecasts for the IP-business were reviewed against the already existing business. The trend in 1999 is that most larger ASIC designs include Virtual Components (IP) and that the IP business starts to take off. This was completed by an outlook covering present and upcoming research activities in the reuse field.

2 EVALUATION OF TECHNOLOGY AND THE MEDEA DESIGN AUTOMATION ROADMAP

Joseph Borel[*], Anton Sauer[**] and Ralf Seepold[***]

[*]STMicroelectronics, Crolles, France,
[**]MEDEA, France,
[***]FZI Karlsruhe, Germany

2.1 INTRODUCTION

The semiconductor industry has been growing at an unprecedented rate since its start in the early 1960s. It capitalized on the outstanding properties of silicon and its stable oxide, which allowed the advent of the CMOS (complementary metal oxide semiconductor) process, the leading process for the whole semiconductor industry. An average growth of 15 to 16% per year in semiconductor sales imposes formidable challenges in terms of the huge investments needed for manufacturing.

A rapid return on investment through the speedy design, production and commercialisation of innovative, advanced products (in the latest available processes) with high added value at systems level is mandatory. This requires new methods to design such complex systems, especially if the system is integrated on a single chip, mixing several functionalities.

The MEDEA Design Automation Roadmap [Mede00] is a forecast of how design automation of semiconductor manufacture could evolve in Europe, based on the current status of a lack of local industrial development but a tremendous reservoir of knowledge. A recent study has shown what are the European strengths, mostly on system design and what are the centres of excellence in this area [Eco96]. No additional comments will be done in the present document. This would make it possible to pave the road to new design solutions and give true capability to influence US developments in the design automation markets, as is evidenced by frequent technology partnerships with US software vendors. Recently, a significant increase has been seen in European start-up companies active in advanced design automation domains, such as hardware/software co-design, IP reuse and deep submicron effects.

The present document is not in competition with earlier roadmaps [Edaa97] but is believed to be a more in depth analysis of what the present areas are in which the IC industry should drive progress for early access of silicon products, mostly in the SOC area.

However it should be recognized that the evolution to complete silicon SOC capabilities has been mostly driven through silicon process evolution capabilities - an enabler at the same level as design automation, the end product being the silicon chip itself that sells on the market. The early accessibility of products in the latest silicon is critical for maturity of these processes. And Europe is lagging behind the USA, although process development is often well synchronized. One strategic European objective should be to engineer design solutions more rapidly, in various silicon application platforms of choice, to be able to develop early reusable system IPs for the next generation of products. These system IPs will include basic digital functions but also value-added functionalities such as analogue (traditionally a recognized European expertise), radio frequency, embedded memories (SRAM, DRAM, non volatile), and micro mechanical functions.

2.2 TARGETS OF EDA ACTIVITIES

Over the past 30 years, the semiconductor industry has been growing so fast that 'process' maturity has scarcely been achieved. Process manufacturing is highly efficient and predictable but design automation lacks maturity in its engineering and performance. This is a major bottleneck, preventing exploitation of the full possibilities of the silicon process available today. The Design Automation Roadmap gives targets, appropriate to Europe, to help bridge this gap.

Major efforts are required in four main areas:

1. Consolidation of design automation technology in areas such as IP (intellectual property) system blocks for greater system knowledge reuse, hardware/software co-design and deep submicron back-end and verification (60 to 70% of design time),
2. Setting up of system-level design solutions, including virtual silicon capabilities for early prototyping and system specification validation,
3. Targeting one-month design cycles in silicon system platform solutions covering focussed markets in which most of the software and IP developments can be reused,
4. In-depth engineering effort for design automation solutions based on what has been already achieved in silicon processing.

It is believed that Europe can take up such a challenge thanks to:
- An existing tradition of co-operation between the different players;
- A high level of maturity in silicon production; and
- A deep knowledge and expertise in design automation issues.

Work on implementing the roadmap should lead - if properly staffed - to a significantly better competitive position of the European semiconductor and systems industries. It will provide the competitive advantage of higher system added value in the silicon processing area.

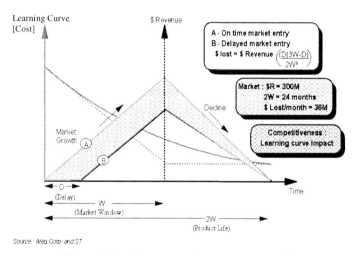

Figure 2.1 Time to market revenue penalties

There are several dimensions that need to be reassessed if Europe wants to improve design automation for semiconductor production. The most crucial for SOC products is time to market (see Figure 2.1). The winners in the market will be those organizations able to carry out full concurrent engineering for both process and design automation - including libraries and IPs [Bore97]. Electronics systems houses require from their EDA solution suppliers a hardware/software co-design solution, a fully qualified library and IP catalogue - including core, SRAM (static random access memory), DRAM (dynamic random access memory), analogue blocks and RF (radio frequency) functions - and a short back-end implementation time.

Figure 2.2 shows the basic functionalities for concurrent development in the design of SOC products in new emerging processes. The evolution of complexity and the need to optimize SOC for high performance applications requires to derive from the ASIC technology platform dedicated platforms for these critical applications (e.g. the mobile wireless applications need RF, very low power, antenna amplifiers which are not available in an ASIC platform).

16   VIRTUAL COMPONENTS DESIGN AND REUSE

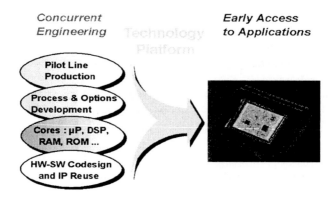

Figure 2.2 Technology platform

2.3 THE SEMICONDUCTOR INDUSTRY CHALLENGE

The main challenge in the semiconductor industry is to bridge the gap between technology and design automation capabilities. There are two main domains to consider:
- Developing basic design automation technology to cover design feasibility up to the specification and the reuse of as much as possible of what is presently available (libraries, macros, specific IPs...).
- Optimizing highly automated specific design solutions to reach a breakthrough in design cycle time, thanks to a dedicated solution for each major silicon application platform (multimedia, wireless...) [Claa99].

2.3.1 Basic design automation technology

By design automation we mean both fully automated parts of a flow or more interactive ones that allows more intervention from the designer.

The most urgent task in design automation is to have a full set of general tools and methodologies covering the needs for designing integrated systems-on-chip. This will be achieved by bridging the process-design silicon gap as seen in Figure 2.3. To achieve this aim, developments must be carried out concurrently in the following areas:
- IP reuse;
- Hardware/software co-design;
- System level design; and
- Back end with deep submicron effects (cross talk, thermal effects...).

IP reuse is mostly advancing along the right lines with the standardization work carried out by the VSIA (Virtual Socket Interface Alliance) group [VSIA00b]. But the most important effort must be made internally within the companies concerned. They need to define their own ways of making it happen quickly, and this requires a huge effort in culture change as well as re-engineering of existing company IPs and methodologies for new IP creation in a high quality environment.

Hardware/software co-design has been the scene for one of the major recent efforts to develop software products in co-operation between design teams and software vendors and will see solutions arriving on the market in the time frame 2000 to 2002.

System level design will target early internal silicon solutions (specification validation with the customer) and links to system level design flows down to the hardware/software co-design flow mentioned above.

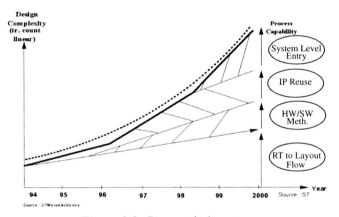

Figure 2.3 Process design gap

2.3.2 Silicon application platform solutions

In areas of high market growth which require full use of silicon capabilities (in terms of performance and time to market), dedicated platform solutions will be necessary to reach breakthroughs in

- Designing complete SOC products in months rather than years; and
- Approaching the intrinsic computational efficiency (ICE) of silicon.

One-month system design from specification target can be reached only through a silicon system platform concept in which all the basic libraries and cores for this application are existing and validated on silicon (as part of the general technology platform, covering various performances goals such as low power and high speed). Thanks to optimized process architectures for applications and pre-verified application-defined peripherals, customized application software can be developed very quickly and put on silicon.

As processor architecture and peripherals are mapped on applications, a better use of the intrinsic computational efficiency of silicon is expected. It represents the evolution of the present microprocessors computational efficiency CE (in MOPS/Watt) versus the intrinsic computational efficiency (ICE) of silicon [Claas99].

Computational efficiency is defined as 'How much we can get in terms of computer power per Watt' and is highly dependent on architecture, algorithm choice, design style and silicon implementation, matching with the application. The gap between ICE and CE is increasing with the current design approach (standard microprocessors), and potential improvements of 10 to 100 times can be expected. This will be a key differentiation in SOC optimization relative to application and to process capabilities (speed versus power consumption, versus cost), together with time to market and first silicon success.

2.3.3 Design automation engineering

It is common sense that silicon process is highly engineered; moving from one process to the next will introduce a limited number of completely new techniques (or equipment). And it is an evolutionary process where the measure of the output of the system is kept under control. This is typical of a system where operations have measured outputs (e.g. the number of wafers / the number of operators x hours in a fabrication facility) and which allows for optimization and high efficiency.

The situation is different in design automation, following a more revolutionary evolution where the 'processes' are less mature and change significantly from one generation to the next. In this situation, constant improvement is more difficult [Kaiz]. We must address the major challenges of doing things faster and with more control as in a kind of accelerated 'Innovation + Maintenance' cycle - Figure 2.4.

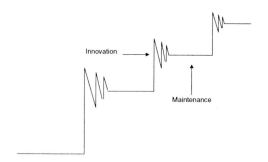

Figure 2.4 Innovation and maintenance cycle

To be able to control innovation and quality results, a set of reasonable parameters in the flow of design automation must be chosen and these parameters must be carefully defined (e.g. equivalent transistor / human x day). This is not a simple task when mixing analogue, RF, digital and memory on a single chip. When the measurement system is agreed, it will have to be checked on practical examples to build good confidence in the results and calibrate the method.

As a consequence, the evolution of the maturity of design should be significantly improved thanks to these quantitative evolutions. There is no doubt that we must find a way to accelerate design automation maturity in a more constrained manner than in the past - it has been engineered since the 1990s, but at a slow pace. This is also a major element in bridging the gap between design automation and silicon processing, reaching an earlier maturity than shown by 'classical IC products design' in Figure 2.5.

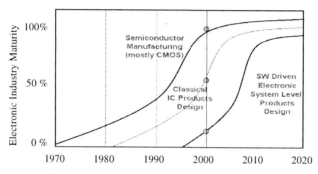

Figure 2.5 Classical IC products design

2.4 ROADMAP COVERAGE

The roadmap addresses the topics mentioned above and will give a targeted evolution until 2005 where we still have a good view of what can be influenced (after every yearly review there will be a shift of one year in the time window).

In this roadmap, European specifications are underlined in terms of:

- Strong analogue-digital culture (with RF incorporated);
- System level knowledge specifics shared from past European co-operative programmes;
- IPs as an urgent and mandatory approach to creating a step change in design efficiency increase and to be integrated in the silicon system design platforms at higher levels of complexity;
- Back-end as an engineered flow integrating deep submicron effects (cross talk...), and
- Top-down verification starting from system description.

20 VIRTUAL COMPONENTS DESIGN AND REUSE

The general architecture of the mixed analogue/digital design and test flow is shown in Figure 2.6. This shows the backbone of the digital flow from specifications validation with the customer until final packaging.

Additionally are represented:

- The analogue and mixed analogue digital flow on one side;
- The IP reuse and the test solutions on the other side.

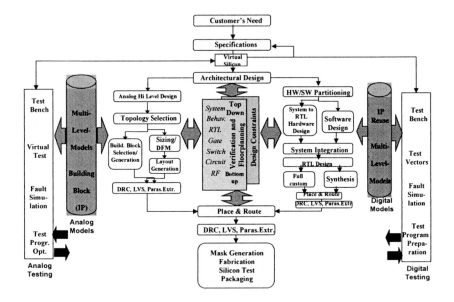

Figure 2.6 Analogue/digital design and test flow

All the above steps are to be considered in an efficient verification environment at the level of the complexity of the problem: from higher description levels (C, C++) down to single-gate behavior. They must be properly selected for a range of applications to get more optimized products.

To illustrate the roadmap, the following definitions are considered:

- "Research" - this covers the early appearance of a concept and demonstration of its feasibility on demonstrators;
- "Development" - this means that research results are believed to be applicable to solve problems, as breakthroughs in the evolution of electronic design automation capabilities and
- "Quality/production" - this ensures that the solution is production proof and has been fully debugged and can be supported.

MEDEA DESIGN AUTOMATION ROADMAP

For sake of simplicity, the roadmaps will only show breakthroughs and duration's to solve these breakthroughs in terms of R/D and quality/production as defined above. Accordingly milestones in terms of results during these periods will correspond to:

- Short term release - covers the appearance of a new concept;
- Medium term release - first results on demonstrator; and
- Generic applications - production solutions availability.

The most relevant breakthroughs of the detailed roadmap are shown in Figure 2.7 with forecast time frames for appearance: the major topics are listed vertically in much the same order as they would appear in a design flow. The horizontal axis gives the maturity time frame of the solutions as seen today. New content will appear when bottlenecks show up and will be introduced when and where necessary.

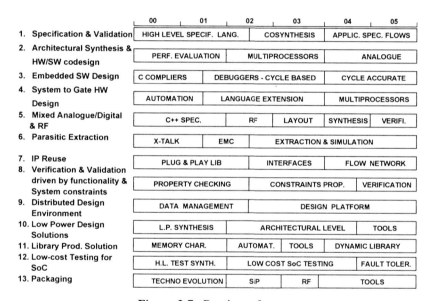

Figure 2.7 Roadmap forecast

The topics selected correspond to the present priorities as seen by the designers and could change later on due to the evidence of new bottlenecks. A five year time frame (moving window) has been chosen to underline the application drive of the roadmap.

All these tools and techniques should be developed internally by European companies - systems or semiconductor - or be developed in co-operation with major software vendors. These solutions will improve the competitive position of the European electronics industry. They can lead to the creation of new start-ups in these strategic areas as this has already been shown in the MEDEA programme.

In the following we focus on chapter V.7 IP reuse of the MEDEA Design Automation Roadmap, because we believe IP reuse has a short term return on investment, improves design quality and time to market and impacts most of all the other areas of EDA.

2.5 IP REUSE ROADMAP

2.5.1 Status

Today's reuse is either restricted to dedicated solutions that are integrated into specialized EDA tools or takes place in parallel to the conventional design flow. The introduction of reuse-oriented techniques requires more than a pure technical enhancement of the tools or the flow but the introduction of business models customized to the companies' needs.

With respect to the challenge to invent a complete solution, current approaches are mostly restricted to certain levels of abstraction, technologies, design flows, tools or even to a different business model that the one applied. Efficient reuse has to follow a four dimensional approach this is facing hardware, software, digital and analogue (mixed-signal) design. None of the available tools can manage these dimensions, and furthermore, none of the existing tools can handle the complexity of the next or even the current chip generation.

Today, large and complex designs are accomplished in a tedious manner by using a variety of commercial and point tools along with the needed software patches. Such sizes and complexities of designs have been empowered by the vast strides taken in the manufacturing technology of integrated circuits. In order to keep pace with the advances in technology, CAD tools and methodologies are needed to increase the overall design productivity and to reduce the time to market by orders of magnitude. However, there is already a capacious gap between what can be put into silicon and the capabilities of the CAD tools. This gap is going to widen further, and there are two pivotal solutions to combat this problem:

- Design reuse in conjunction with a migration to higher levels of abstraction will augment the productivity by a factor of 5 to 10 after the turn of the millennium, and
- Prototyping will decrement the time to market.

2.5.2 Goals

The introduction of reuse is a complex task, because several elements of the current design flow are modified. In contrast to the introduction of a tool, comprehensive reuse requires a new design methodology and a modification of the design flow. Since a general design methodology must be customized to the specific needs of a design environment, it is not possible to predict a unique schema for successful reuse introduction.

Commonly, the successful application of reuse methodologies is associated to terms like higher design quality and increased productivity. Due to the complexity of both the required methodology and the individual design environment, the results achieved by reuse can be classified with regard to their time dependency. That means, the benefit of reuse is closely related to the time that has been spent to prepare the reuse. Besides evaluation techniques, which provide models and methods to determine the benefit of reuse, short term objectives and long term objectives can be identified. A short cut out of short objectives is presented in the following list:

- Complex design

    Design reuse methodologies provide efficient methods to support the management of complex designs, because the top-down design methodology and derivation relations are maintained. These meta data are provided for future application development.

- Concurrent engineering

    Since the introduction of design frameworks, concurrent engineering can be controlled and customized for individual design environments. Design reuse and concurrent engineering use the same meta data schema and domain analysis techniques. A unified application of design reuse and concurrent engineering maintains the benefits of both methodologies.

- Short time to market

    The potential of reuse applied to current designs can be efficiently reused for future designs. Since about 90% of redesign data are based on reuse, future product development time and time to market are significantly decreased.

- Design efficiency

    Due to the fact that the most innovative potential and know-how is contained in designs, which have been developed and stored in internal libraries, a structured reuse of ideas will influence the possibility to create an innovative design reuse.

The short term objectives of IP reuse are manifold since the main stream activities will benefit from related activities and investigations. Design for reuse and reuse of design strategies are in the focus of IP reuse, since fundamental multiplicators will be reuse database development, management, standardization, interface design and inter-company strategies that are required to implement mature reuse management systems (RMS).

In a short term period, the development of highly efficient reuse-oriented tools will be focused on more than one of the mentioned four dimensions: hardware, software, digital or analogue (mixed-signal) reuse. In spite of the fact that the development will take place on separate sites, a fundamental base has to be invented in order to install closely related links. This base will serve as a starting point to develop tools and systems that can manage not only one but up to four dimensions. A first step into this direction has been realized by inventing interfaces to interconnect virtual components, like it is proposed by the Virtual Socket Interface Alliance [VSIA00b]. At least tools have to follow a certain standard to offer both the exchange of virtual components and the exchange of the best suited tool in a comprehensive reuse management system. Furthermore, with respect to already existing reuse database classification and query techniques, more than one dimension (currently the digital domain) have to be integrated.

The transfer of know-how must not lead to a stagnation of the development and enhancement of methodologies and tools which are currently under development in the digital domain. Several techniques, like e.g., classification, similarity detection and evaluation, databases or object-oriented techniques must be improved to build up mature support for reuse-oriented design that still has to cope with both design of reusable/virtual components and reuse of existing (virtual) components by taking protection mechanisms into account.

Since first successful digital reuse methodologies have been invented and derived from software reuse, analogue reuse will benefit from several digital reuse strategies that could be partly transferred and customized to different environments. This bottom-up approach must be synchronized to a top-down approach leading to the same objective, and therefore, support its own evaluation.

The overall design goal of bringing a product to the market in the smallest possible time is the main motivation for introducing reuse into development. Regarding natural boundary conditions for development like fixed budgets, a significant reduction in design costs is directly related to efficient application of reuse methodologies. As a consequence, reuse must not be restricted to certain levels or the complexity of components. The innovative gain that is expected from this methodology is only possible if horizontal and vertical reuse are addressed. Of course, this does not exclude an additional application of the described conventional approach that can also be seen as a complementary method. The benefit of the reuse process is larger if it is applied very early in the design phase.

The long term objectives of comprehensive solutions are dedicated to the already mentioned four dimensions. But the task of integration is highly complex and it requires that it can be derived from fulfilled short term goals. The maturity of a comprehensive solution offers support to customize a design flow to the specific needs of a company while providing an independent platform for incorporating tools that are compliant to industry and/or de facto standards. Hardware/Software Co-design with digital, analogue and mixed-signal components all integrated on a single system-on-chip design must be handled by a comprehensive reuse management system. As mentioned earlier, design reuse, rapid prototyping and verification approaches must be merged into a common model. The long term objectives of comprehensive reuse systems must offer a mature platform to separately incorporate applicable point tools like embedded reuse-oriented strategies (e.g., hierarchical synthesis) as well as a platform for plug&play of efficient reuse support tools, like for example a complete reuse database. The sketched out scenario requires intensive application- oriented research that is guided by industrial applications and enriched by research results. These are not available yet but are required to reach the overall objective of a common customizable reuse management system that enhances design quality, productivity and predictability.

2.5.3 Proposed work

A group of experts led by R. Seepold have identified necessary future work in the following fields (Figure 2.8):

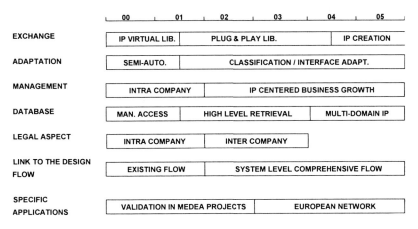

Figure 2.8 Roadmap for IP reuse

2.5.3.1 Exchange

- Needs versus process evolution:

  The exchange of virtual components is a prerequisite for reuse but it is often a tedious and time consuming task to design a component for exchange or to make a component exchangeable. With respect to common design flows, exchange of components is missing because the return on investment has not been taken into account when a reuse management system or the methodology for IP qualification is missing.

- Status:

  Currently two kinds of IP are available on the market: qualified IP (often referred to as Virtual Component, or VC) and unqualified IP. Unqualified IP means that there is no guarantee of plug and play in other environments than in the environment the IP has been designed for. Qualified IP describes IP that is contained in a virtual IP library and that is applicable in different environments [Seep99a]. Typically, qualified IP has been installed between system manufactures who are building a system-on-chip design and a specific silicon vendor. Qualified IP is also exchanged among silicon vendors. Currently, unqualified IP is exchanged between independent R&D companies and system manufactures and different silicon vendors. Besides this fact, software reuse is becoming more important because its flexibility supports the application of reusable components by taking into account that different environments may appear.

- Short-term goals:

To overcome the barriers of tools, flows and methodologies, an IP virtual library must be installed to ensure the trade of Intellectual Property: It also has to certify that a certain IP from certain source meets the necessary criteria for plug and play in target designs. Multiple organizations could provide the virtual library function and they could act as an IP-Qualification centre. In the short term, this library could be installed in-house, supporting the in-house interfaces and guidelines for reuse. The measurement of IP quality is crucial in the areas of implementation or silicon and design (design rules).

- Future needs:

An IP virtual library format has to support very different components and exchange formats: Hardware/Software Co-design in the digital and analogue (mixed-signal) domain. Besides the pure technical challenge to exchange virtual components, IP qualification tools and IP qualification centres are required to make components exchangeable. Furthermore, legal terms are becoming more and more important. The in-house strategy has to conform to de facto standards to install an external IP virtual library for inter-company exchange, and therefore to overcome intra-company reuse.

2.5.3.2 Adaptation

- Needs versus process evolution:

The evolution from a pure ASIC design to complex ASICs with few IPs and finally to a system-on-chip design with plug and play of components requires a *design for reuse* strategy to install the required parameterization capability.

- Status:

Different methodologies are available to deal with the semi automatic adaptation of virtual components. At least a split of pre-condition, core and post-condition has to be performed when a component is designed for reuse. A component can be described at different levels of abstraction supporting black-box, gray-box or white-box reuse. In the final stage, the design of more complex components can be derived from already existing components that are stored in a reuse database. Other techniques are supporting object-oriented methodologies to apply its paradigms derived from the software reuse domain.

- Short-term goals:

  In the short term, successful reuse is influenced by the availability of mature methodologies to ensure an efficient adaptation of virtual components to the environment in which they should be integrated. Especially in this area, the complexity of the components requires major research activities to ensure that component adaptation can be performed quicker and with lower effort than designing it from scratch. Parameterizable components are especially well suited for reuse. The different parameter settings lead to a large amount of different configurations, i.e., to a different functional behavior. This poses a major challenge for the validation of the components as the simulation effort is intractable. On the other hand, the correct functioning has to be guaranteed for all admissible parameter settings to be able to qualify a component for reuse. New solutions have to be found and applied, e.g. formal verification. Virtual components should provide the flexibility to cut of pieces, and while following a general design for reuse flow, adaptation flexibility must be part of this flow.

- Future needs:

  In a second step, the adaptation of components should be hidden from the reuser since the complexity requires many information details that are beyond the scope of the designer who wants to instantiate and adapt the component for a certain design. These mechanisms must conform to available classification techniques that compute the similarity of a component to a given specification. And therefore, adaptation is closely related to component exchange and database management.

2.5.3.3 *Management*

- Needs versus process evolution:

  Due to the tremendous increase of complexity of system-on-chip, the integration of these systems also becomes more and more complex. If one has to integrate a system, the designer has to deal with components like processor cores, DSPs, communication cores, dedicated hardware, and of course software blocks. It must be decided which part of the functionality should be implemented on which component of the architecture. Designers need tools to aid them in the process of system integration.

- Status:

  Isolated reuse has been overcome in some companies. The status is that intra-company reuse strategies are being implemented.

- Short-term goals:

  After the installation of in-house solutions, there is a possibility available in which third-party vendors might offer components that are compliant to the internally used technology and flow. Via the qualification of virtual component libraries, independent R&D vendors must get a platform to introduce components that are acceptable by system developer or silicon vendors. In order to open the gate, several mature quality measurement metrics must be applied to guarantee both the quality of the virtual component and the licensing of the IP. In a short term, the management structure has to reflect the current status of the installed solution. An IP life-cycle model containing all required parameters should be provided to support the different phases like for example "garbage administration" facilities. Basic functionality must be implemented in the reuse database (cf. Section 2.5.3.4)

- Future needs:

On the long term, a set of tools should be customized to the reuser's needs to evaluate the quality and reliability of the virtual component. Therefore, new business models have to be developed inside and outside the different market participants. The main stream development must be closely related to other IP techniques to be compliant to technical and management (e.g., legal) aspects.

2.5.3.4 Database

- Needs versus process evolution:

Databases for storing IP have to overcome pure storage mechanisms of IP or versioning, and therefore, to fulfil at least two more tasks: efficient storage of IP and offering customizable techniques for IP retrieval like classification, similarity metrics, support of component adaptation and IP exchange, etc.

- Status:

The development of database and storage applications dedicated to IP reuse is available on file structures and commercial products. Often, companies prefer relational applications, because these types of databases are already installed for other kinds of storage. Depending on the technologies of the system that is supporting IP reuse, also other types like object-oriented seem to efficiently support the internal data structure of the reuse management system.

- Short-term goals:

On a short term basis, the development of database support will be restricted to existing in- house database technologies and systems. An internal data model of the IP reuse data should be modelled with the support of implementation independent methods (e.g., OMT or UML) to ensure both the adaptation of the structure to upcoming standards and the transfer into new and innovative database techniques while maintaining the design data and the access. Classification and intelligent metrics (e.g., similarity metrics) must become mature and customizable. The availability of quality measurement tools will become mandatory. IP protection mechanisms are urgently requested to further support the IP market.

- Future needs:

    Since the know-how of the companies is related to a certain type of database (e.g., relational), it seems to be cost intensive to introduce a different type only for IP reuse. Therefore, it seems appropriate to define an independent interface to support different types independent for certain vendors and types. This vision would help to maintain the data for inter-company reuse because this type of reuse must not be restricted to virtual components that are extracted from the database. The database support tools must reach a higher level of maturity that incorporates the four dimensions mentioned (hardware, software, digital and (mixed-signal) analogue).

2.5.3.5 *Legal and business aspects*

- Needs versus process evolution:

    If design reuse is to be restricted to reuse of intellectual property designed in-house, there is little impact on business models and legal processes. However, design reuse methodologies are also driven by the need to externally source a large part of the IP, for three key reasons:

    The shift towards the consumerisation of electronics tends to progressively blur the long established boundaries between applications. As a consequence of the convergence of applications into single products, organizations which traditionally focused on single vertical markets will require to source critical functions externally.

    Companies will have to source IP blocks from outside their own organization, since no single company or division will be capable of conducting a complete design in-house, due to gaps in their internal IP portfolio.

    Finally, time-to-market pressure and the resulting need for shorter design cycles, will also force companies to look for external IP.

    The 3rd party IP market is currently limited by a lack of standard business and legal practices, widely accepted within the industry. Addressing the legal and business issues will allow to create a structure 3rd part market that will serve the overall system-on-chip industry.

- Status:

    Reuse methodologies are restricted to technical approaches and to find solutions on a technical basis. When independent R&D companies try to enter the market offering cores on the web, they are facing the same problem as companies that are going to be an IP provider which is in the moment not their main business. The first available catalogues support the idea of the user browsing to the best suited core but they do not support the potential customer in offering a contract or the conditions of a contract.

    From a business point of view, the lack of best practices lead companies to treat each transaction as unique, devoting time and resource to identification and mitigation of the risks present in each such transaction. For instance, issues like support, documentation or warranty terms will vary widely between different providers, and will be hard to compare. It results a lack of predictability that leads to an increase in risk and cost, ultimately stifling innovation in the VC market and heightening barriers to entry.

- Short-term goals:

In-house strategies and IP protection mechanisms must be integrated into one comprehensive business model supporting IP reuse and incorporating the needs of the most important stakeholders: designer, provider and user. Besides that close contacts to existing initiatives are strongly requested.

Short term, the market will need a commonly accepted definition of the minimum information requirement for a VC to be considered "qualified". The definition could be based on the VSIA's Virtual Component Transfer (VCT) DWG and will require to be consistent with the effort of other bodies that are setting technical standards for VCs. At a later stage, it is envisioned that organizations will offer compliance checking services allowing to perform a technical and business evaluation of a VC, and therefore to enable accurate prediction of the time, effort and cost require to integrate the VC into a system.

From a transaction point of view, the market needs a commonly accepted set of clauses reflecting the current and future state of the VC marketplace. Conventions on issues such as support, scope of licenses, ownership of IPR in modifications and other key issues have to emerge. The first published attempt for defining these clause is the VCX's, that expect to deliver its first version of the "Contract Configurator" in the first half of 2000 [VCX00].

- Future needs:

Upcoming models must be evaluated and quickly integrated into companies' models to ensure the compliance to international regulations and a successful European competition on a fast growing market. On a longer term view, and assuming a development of a 3rd party IP market, we could envision organizations to offer royalty collection, tracking and auditing services like "clearinghouse" services are offered in other marketplace (e.g. music industry).

2.5.3.6 *Link with Design Flow*

- Needs versus process evolution:

The link to the design flow depends on the acceptance of the designer to apply reuse and the capability of the tools to be customizable to the requirements. Since IP reuse is at its beginning, point tools are available in the flow but often they are not portable to different flows.

- Status:

Initial initiatives for design reuse have shown that the availability of reuse support can be installed in parallel to a design flow but the efficient support of IP reuse cannot be reached without the installation of closely related design flow links. Current developments have shown that IP reuse must be integrated into the design flow. This integration does imply that the design flow has to be modified and extended. A pure parallel solution will neither be accepted by the designer nor the expected support for plug and play of virtual components can be reached.

- Short-term goals:

In a short term, the design flow must be extended to include IP reuse tools and methodologies. This requires a very flexible customization capability and the introduction of standards. This is true for all dimensions mentioned.

- Future needs:

Since virtual components are developed to be exchangeable, the design flow must follow this capability to offer both comprehensive IP data that are required for reuse and the capability to link tools and data (components) to different design flows with tools from different vendors.

2.5.3.7 Specific Applications

- Needs versus process evolution:

The needs of specific tools and applications are synchronized with several running projects but it has to be ensured that mature concepts will be made available, and furthermore that a future market incorporates upcoming technologies. This can be reached if both research and development are closely connected to face the same objective.

- Status:

Specific applications are required for the core competencies of European industry, like e.g., telecommunication, automotive and multimedia. Dedicated tools are under development.

- Short-term goals:

The complexity of the IP reuse domain and the broadening of this topic into other domains, like from digital to the analogue domain, requires that dedicated tools must be developed before a comprehensive solution can be offered. In parallel, research activities should support and ensure the capability to derive the results received from tool applications. However, the transfer of know-how is a moving target because partly established reuse domains like digital reuse have to be developed to become mature.

- Future needs:

Besides the protection of European core competencies on the world market, other innovative approaches like micromechanical environments will become important for IP reuse platforms. With the help of the experience in the mature IP reuse domains, a transfer of know-how has to take place on a methodological level, in order to ensure that new technologies can benefit from already installed know-how without neglecting the development of core domains (e.g., digital design).

2.6 CONCLUSION

The roadmap described in the MEDEA Design Automation Roadmap document has the main objective of achieving better use of the silicon process complexity and intrinsic computational efficiency in a shorter time-to-market through the concept of silicon platform design solutions. They could lead to:
- Better use of silicon capabilities;
- Better optimized systems; and
- Shorter time to market.

This is feasible only in a continuation of MEDEA where the above goals are fully understood and shared by the European community (semiconductor and system houses, universities, engineers and researchers themselves). The result will be more added value on European silicon systems, continuing the successful track records of JESSI and MEDEA.

3 PRODUCTIVITY IN VC REUSE: LINKING SOC PLATFORMS TO ABSTRACT SYSTEMS DESIGN METHODOLOGY

Grant Martin

Cadence Design Systems
System Level Design Group
San Jose, California, USA

3.1 OVERVIEW

This chapter links two trends in virtual component (VC) reuse, and systems design:

1. system-on-chip (SOC) integration platforms, and
2. new methodologies for abstract systems design.

The real power of these techniques derives from linking them together to offer VC re-use at the highest level of system abstraction: this increases design productivity and reduces design risk for SOC beyond that which could be achieved with one of them alone.

Application-oriented SOC integration platforms are a mechanism for the rapid design of integrated circuits through re-use of HW-SW architectures and virtual components. Platforms offer a stable framework into which pre-qualified HW and SW IP *suited to a particular range of applications* can be placed. With an integration platform, rapid creation of design derivatives within the field of use of the platform is possible.

The new systems design approach, function-architecture co-design, models embedded systems in two orthogonal dimensions:

1. an implementation-independent functional model, and
2. an abstract architecture incorporating significant VC's.

These two models are linked via an explicit 'mapping', which assigns the functional blocks to particular processing and communications resources in the architectural model. Performance analysis of the resulting design allows alternative architectures, partitionings and choices of VC's to be explored.

We apply this co-design approach to the problem of integration platform definition for a family of derivative multimedia products. Modelling a reference design application with the platform gives a working example to serve as a starting point for derivative design. Finally, an example of a derivative design is illustrated.

3.2 THE NATURE OF SYSTEM DESIGN

We discuss the linkage of two fundamental trends in virtual component design and reuse, and systems design methodology: the emergence of system-on-chip (SOC) integration platforms as a vehicle for high productivity in SOC design; and the emergence of new methodologies for systems design which operate at an abstract level. As two orthogonal trends in design methodology, these developments offer substantial improvement in design productivity. The real power of both of these techniques, however, comes when they are linked together to offer VC re-use at the highest level of system abstraction. The combination of these techniques increases design productivity and reduces design risk for SOC beyond that which could be achieved with one of them alone.

The latest International Technology Roadmap for Semiconductors has identified SOC platforms as a key trend:

"For primarily cost and time-to-market reasons, we expect that product families will be developed around specific SoC architectures and that many of these SoC designs will be customized for their target markets by programming the part (using software, FPGA, Flash, and others). This category of SoC is referred to as a programmable platform." ([SIA99], p. 6).

At the same time, we see the emergence of system level design as the only credible entry point for complex SOC devices. Entry at the RTL and C implementation levels for hardware and software does not allow sufficient latitude and flexibility to designers to explore alternative architectures, and choices of virtual components. A system level design approach does.

3.3 SOC INTEGRATION PLATFORMS

SOC integration platforms ([Mart98a] [Quin99] [Chan99]) have emerged in recent years as a vehicle for rapid design of application-oriented integrated circuits through the maximum exploitation of re-used virtual components.

The idea of an integration platform is to create a stable framework into which pre-qualified IP suited to a particular range of applications can be placed. With an integration platform approach, rapid creation of design derivatives for specific products within the overall application orientation of the platform is possible.

An integration platform for SOC design is a high productivity design environment which specifically targets a product application domain, and which is based on a VC reuse, mix and match design methodology. The application domain is selected based on market objectives and is focused to allow a high probability of reuse over a target period of time. The integration platform consists of:

- An SOC integration architectural specification describing, on the hardware side, the on-chip bus, power, clocking and test architectures, the IO configuration, the substrate isolation design, and performance, power and area constraints on VC blocks,

- and on the software side, the base software architecture, including RTOS, task scheduling, inter-task communication, device drivers and other SW-HW communications, inter-processor communications, and the layering of middleware and applications software.

- A portfolio of virtual components (pre-characterized, often through fabrication, pre-verified, pre-qualified, pre-conditioned VC blocks) pre-staged on the target technology to meet the constraint ranges specified in the integration architecture. In general, at least some of these virtual components will be programmable cores (a more generic layer), and at least some of them will provide significant differentiating value for the range of products to which the integration platform applies (the more application-specific layer).

- A set of proven, documented, VSI-compliant block authoring and chip integration design methodologies consistent with the models used for the VCs and scripted for very rapid assembly and verification of the SOC device.

- Guidelines for block design dealing with mixed signal, Design For Test, Design For Manufacturability, and yield issues.

- A design verification methodology and environment with appropriate models, which meets the HW/SW co-verification requirements of the application domain. This could include system environmental models, rapid prototype models, bonded out cores, behavioral and performance models, bus functional models, and in-circuit emulation tools.

- A reference design illustrating the use of the platform and its VC options to build a specific product example.

The platform approach has been applied especially strongly in wireless handset design, for current standards such as GSM and CDMA, and in upcoming 3rd generation wireless standards. ([Mart98c] [Mart98a]). Leading companies applying this concept include VSLI Technology [Fost98], and more recently, Phillips Semiconductors [Claa99], which purchased VLSI Technology in part for its platform designs and intellectual property.

3.4 FUNCTION-ARCHITECTURE CO-DESIGN

Over the past two years, a new approach for embedded systems design, function-architecture co-design, has been explored as part of the Cadence 'Felix' project ([Mart98b] [Chak99]). A product that has been commercially released from this project is known as VCC^{TM} [Sant00]. In this approach, embedded systems are modelled in two orthogonal dimensions: first, as an architectural and implementation-independent functional model, and secondly as an abstract architectural model of a candidate implementation, often incorporating significant re-used virtual components. These two models are linked via a 'mapping' process, which partitions the functional blocks in the behavioral description into HW and SW realizations and assigns them to particular processing and communications resources in the architectural model. A performance analysis of the resulting mapped model is carried out, and various alternative architectures, partitionings and choices of virtual components can be explored before arriving at an optimal architecture on which to realize the application function. Thus function-architecture co-design is one of the class of 'Y-chart' methodologies [Kien99].

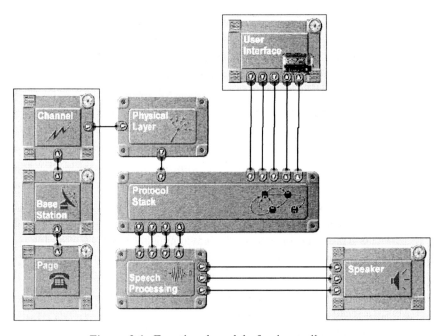

Figure 3.1 Functional model of voice mail pager

The functional model is built as a heterogeneous composition of functional models drawn from a variety of modelling environments, including language-based models (C, C++, SDL), co-design finite state machines (CFSM's), and statically-scheduled dataflow models (SDF), which are linked together using a discrete event semantic with well-defined firing rules.

Figure 3.1 illustrates a compositional functional model for a voice mail pager application. This is built with individual functional block models created in a number of modelling environments:

- A baseband-processing model for the physical layer developed in a dataflow algorithm tool such as SPW or Cossap.
- A protocol stack developed using SDL and code-generated C.
- A user interface model built as a finite state machine and with C.
- A speech-processing algorithm provided as a functional model by a VC vendor.
- A testbench built within the 'Felix' modelling environment.

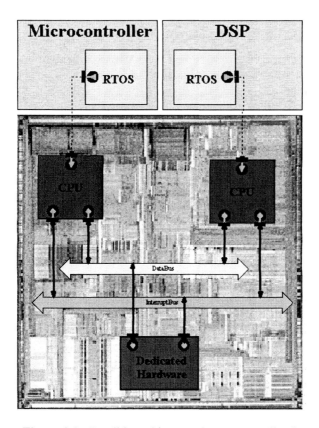

Figure 3.2 Possible architecture for pager application

The architectural models are built using a 'relaxed' modelling style. This avoids details of signals and pins, and links together abstract models of HW and SW components via, for example, delay equations or models for HW blocks, simple scheduling and latency models for Real-Time Operating Systems (RTOS's), and priority and latency models for SW tasks created with SW estimation techniques. Pre-characterized virtual components can be modelled at this abstract level with the appropriate techniques [Mart98d].

Figure 3.2 illustrates a relaxed architecture on which the voice mail pager application behavior will be mapped. It consists of two processors (with their associated RTOS's): a microcontroller, and a DSP; and a dedicated HW block.

The 'Felix' approach separates functional task models from interface models and supports mapping of inter-task and block communications to specific communications resources (HW-HW, HW-SW, SW-SW) at both an abstract frame or packet transmission level and at 'refined' levels: for example, via generic bus protocol transactions modelled with accurate on-chip bus latencies and arbitration. This mapping (see Figure 3.3) is used as the basis for performance analysis of partitionings and assignments of functions and communications to various VC resources, architectures, and choices of VC's. The mapping and analysis can be carried out both at abstract and refined levels, converging on an optimal architecture for implementing the applications behavior.

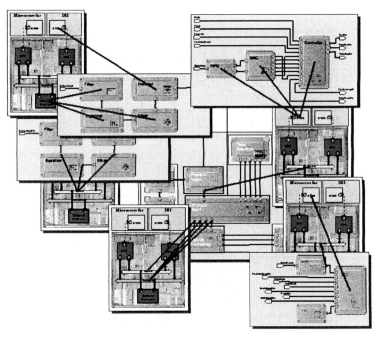

Figure 3.3 Mapping of voice pager behavior to architecture

Finally, both HW and SW implementation-level netlists, virtual component choices, task priority structures, memory maps and other design decisions are exported to HW and SW implementation flows from the 'Felix' toolset.

The 'Felix' system design initiative is reaching a state where partners such as Phillips (via the Esprit COSY project) are applying the methodology and approach to applications such as multimedia systems ([Brun99] [Kent99]). In their work, they used the function-architecture technology as a means to import applications or behavioral models for video systems, described in their own system modelling approach called YAPI (Y-chart Application Programmers Interface). One reason for this was their desire to use a different 'model of computation' than that supported in the Felix co-design technology CFSM; in particular, a lossless parallel process model called Kahn Process Networks. This illustrates two key requirements: the need for co-design technology to support several different computational models, and where this is not available, to support generalized model importation mechanisms.

The application modelled was an SOC decoder for digital video broadcast application, involving about 60 IP blocks, including a MIPS processor and hardware accelerating modules.

The function-architecture co-design technology was thus used as a general vehicle to map applications functions onto a system architecture under continuous refinement, and to explore various mappings and communications mechanisms. To make architectural definition at an early point in the development of the Felix technology, Phillips researchers developed a kind of architectural generator called SYS-BUS that interacted with the co-design technology. Communications latencies were modelled using simple delay equations for token-based transfers. Although individual transactions were expected to differ significantly between the system level of design exploration and a detailed implementation, at the aggregate system level over many transactions, good fidelity between these two levels was expected for video applications.

Benefits of using this approach included simulation speed - for 66 processes and 201 communication channels, one second of real-time video was simulated in less than 5 minutes. This permitted both effective IP selection and validating the algorithmic compliance with customer requirements. The clear separation of function and architecture, and between block function and communications, was an effective approach for defining an SOC integration platform, and allows more system designers to work on derivative products without involving the 'gurus' or experts all the time. Other benefits included modelling fidelity, allowing alternative implementations to be compared, and the ability to create a library of key system level functional IP blocks with alternative implementations, allowing effective future reuse.

3.5 LINKING THE TWO CONCEPTS: PLATFORM-BASED SYSTEM CO-DESIGN

The 'Felix' approach is a powerful one for modelling systems of many types, especially embedded systems, with its attention to modelling of SW tasks and processors. It gains special power if combined with the concept of an integration platform. If an integration platform is modelled using the 'Felix' approach, together with appropriate abstract models of HW and SW IP or virtual components, then the key design decisions involved in derivative design can be taken at a systems level rather than the implementation level. Furthermore, by structuring the libraries and associated system-level models carefully, the derivative product designer can be given a well-planned environment for design in which the choices of virtual components will be only those which are pre-characterized and matched to work in the platform. This can include the 'collaring' or 'wrapping' of virtual components to match the on-chip bus architectures with either bus-specific interfaces, or emerging adaptation interfaces such as the VSI Alliance On-Chip Bus Virtual Component Interface (VCI). [VSIA00a]

Key derivative design decisions that are taken at the system level include:

- Support for required new standards
- Functional bandwidth, throughput, and latency requirements
- Communications resource bandwidth and latency, modelled at abstract and detailed levels of refinement
- Derivative power consumption and cost
- SW task decomposition, messaging and priorities
- How much of the platform reference design can be re-used in the derivative; as-is, or with modifications?
- How much estimated margin is there in the derivative design to handle unexpected issues during implementation?

If we turn the architecture used to implement the Voice Mail Pager example in the previous section (Figure 3.2) into a platform on which to realize multiple derivative designs within the field of use for the platform, we find that the basic architecture allows a considerable range of derivatives. For example, we can map a JPEG encoder/decoder application onto the same basic HW-SW architectures and VC's (Figure 3.4).

However, when we attempt to construct a combined derivative product on the platform, consisting of a Voice Mail Pager with image display capabilities (thus acting as both a Voice mail and image mail pager), using the JPEG decoder behavior (Figure 3.5), we find that an initial mapping of function to the platform architecture results in a corrupted voice page decoding, due to the audio decoding ending too late (see top half of Figure 3.6).

There are several possibilities for fixing this problem in derivative design. One is to move some of the JPEG decoding which caused the stall in audio decoding into dedicated HW (for example, the JPEG IDCT). Another is to re-prioritize the QCELP audio decoding to give it a higher priority. When we try this second option, we obtain both correct audio decoding and correct image decompression (see bottom part of Figure 3.6). This manages to achieve correct derivative product design without the expense of additional dedicated hardware.

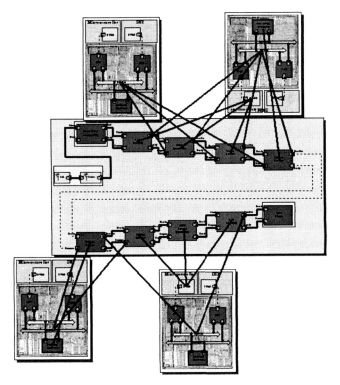

Figure 3.4 JPEG encoder-decoder mapped onto the platform

This example illustrates the power of doing derivative design on a system level model of a platform. Many architectural and mapping alternatives can be explored in much less time and effort than is possible through similar experiments at the RTL-'C' level of abstraction. Furthermore, the types of alternatives that can be explored in the derivative design phase are constrained by the platform architecture and VC libraries so that the process of either converging on a workable system level design, or concluding that a derivative is outside the scope of the platform field of use, can occur much more rapidly and at lower cost than at lower levels of design abstraction.

This allows decisions to be made on alternative design approaches, such as custom hardware design or hand-tuned software implementation, at higher risk and cost, to achieve designs, which go beyond what the platform can support. But these decisions can be made more confidently having explored the scope of the platform first.

When a platform-based derivative design is exported from the 'Felix' system level design methodology, all decisions which have been made at the systems level, regarding virtual component choices, software task priorities, etc., are encapsulated in generated models and design information.

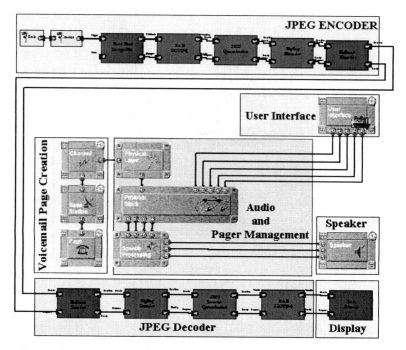

Figure 3.5 JPEG decoder and voice mail pager derivative application

Since the system level libraries and choices of architectural components and structures have been carefully controlled to work with the platform of choice, the implementation processes in HW and SW can be quickly and reliably executed so that a truly rapid (in some cases, only a few weeks) derivative design can be created.

Applying the 'Felix' approach to a platform concept emphasizes the importance of modelling at least one reference application in the environment, to serve as a starting point for derivative designers and as an example of how an application can be modelled and mapped to the platform architectures.

The integration platform concept can be applied to various levels of platform ranging from fairly generic processor, RTOS and bus choices which will apply to a wide variety of application domains, through to an almost-complete hard platform in which the product variation is expressed primarily through derivative SW development. The system design methodology allows designers to work in a more generic through to a more restricted design domain depending on the inherent flexibility of the associated platform. It is easy to see that the risk involved in product design can be controlled via this tailored systems approach.

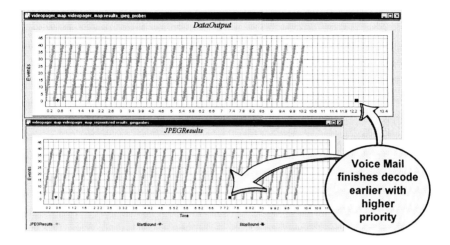

Figure 3.6 Results of two different mappings and prioritizations

3.6 MULTIMEDIA EXAMPLES

The 'Felix' systems design approach has been applied to a basic multimedia platform, and some derivative designs from this platform, as part of an R&D project in SOC design methodology. The basic multimedia platform consists of an ARM7 and DSP Group Oak processor, along with associated AMBA system and peripheral buses. (Figure 3.7).

A reference design implemented on this platform is a 'VOP' (Voice Over Packet, or POTS, or PSTN) - a basic two-way audio system linking two PC-based systems via modem either to the normal telephone network or to a packet-based network, thus providing two-way telephony (Figure 3.8). This is chosen as a simple illustrative reference design for the SOC design methodology. The VOP used G723.1 True Speech audio decoding and encoding algorithms as part of the reference design work.

Another simple application, which has been mapped to the platform (Figure 3.9) is an Audio-On-Demand (AOD) product, a one-way audio decoding system using different decoding algorithms (MP3) than VOP. This was analyzed for performance to ensure correct functioning, prior to design implementation at the RTL and C level.

Although VOP and AOD example designs are simple and well within the scope of the basic platform, the 'Felix'-based methodology has been elaborated as the key front end to a comprehensive SOC design methodology for platform-based design. We have also begun exploring the system design approach in initial platform definition, study and exploration.

44　VIRTUAL COMPONENTS DESIGN AND REUSE

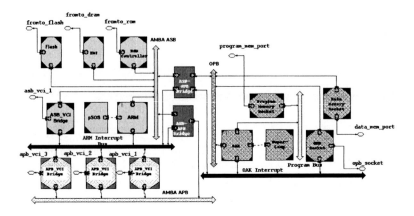

Figure 3.7 Basic multimedia SOC integration platform

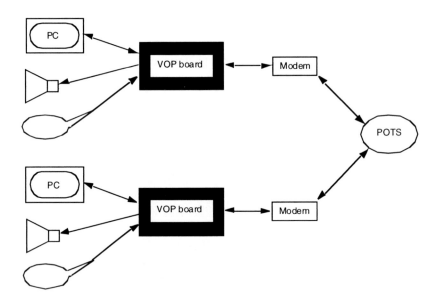

Figure 3.8 VOP application for platform

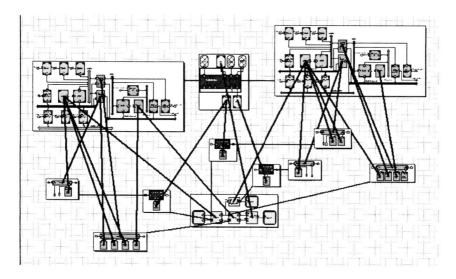

Figure 3.9 Audio on demand application mapped to the platform

3.7 OTHER APPLICATIONS OF PLATFORM-BASED SYSTEM DESIGN

The Berkeley Wireless Research Centre (BWRC) has been using the 'Felix' co-design technology as part of its Two-Chip Intercom (TCI) project [Amme00], to support architectural exploration in the mapping of network protocols and baseband signal processing to software running on Tensilica configurable microprocessors, and dedicated HW blocks, using a SONICS bus as its communications mechanism. The physical layer is modelled in Matlab Simulink and Stateflow, and it is intended to be imported into the co-design technology and combined with models of the network protocol captured in the co-design tool, to form a system model. Exploration of different architectures and component choices has begun, including an analysis of ARM vs. the Tensilica processor, and a variety of mappings.

By using the function-architecture co-design technology for the TCI project allowed up to 4 different architectures and mappings to be explored per day, achieving convergence on a solution within 2-3 days [Raba99]. This followed four person-months of development time, which included learning the technology, deciding on a modelling approach, and modelling the system. This illustrates an important reason why this technology is particularly applicable to platform-based design: the investment in modelling and the learning curve is amortized over the entire stream of derivative designs which can reuse the models of platform architectures, IP blocks and components. Thus the payback with this approach in a platform scenario is many-fold, when compared to a custom chip approach.

3.8 CONCLUSION AND FUTURE

SOC integration platforms and design at the systems level will each continue to evolve. The basic methodologies in these areas have been developed in R&D programmes, and work on linking and applying these concepts to practical SOC design via the platform approach will continue. The key issue now is the deployment and application of these methods to SOC design work in semiconductor and systems design houses. In addition, work on new modelling approaches and requirements, especially for complex heterogeneous systems, will continue. The practical use of these methods requires sophisticated and proven modelling techniques which capture the essence of product behavior without unnecessary detail.

One important area of exploration is the impact on system level design of the highest level of platform abstraction: a soft-programmable platform, whose underlying HW is fixed, and on which application functionality is mapped using embedded software, and reconfigurable hardware blocks (FPGA, LPGA, etc.) for functions which can be optimized through hardware implementation. This has been called a 'Fixed Manufacturing Platform' [Chan99]. The implication of this soft-programmability is that the system level design approach will increasingly resemble a software creation approach, and thus it will be important to study those system level specification and implementation techniques which are emerging in real-time embedded software.

One recent trend of considerable potential for the future is that of Object-Oriented Analysis, as represented by use of the Object Management Group's Unified Modelling Language (UML) [Booc98], to be extended to embedded real-time software [Seli99]. OMG has issued a Request for Proposals for extending UML via a profile covering aspects of scheduling, performance analysis and time, all vital for real-time software [OMG99]. This RFP is expected to attract proposals by mid-2000, and development of associated tools will occur until at least 2001. It is important for the embedded real-time software world and the hardware-oriented system design tool world to converge in order to facilitate the proper design of embedded real-time hardware-software systems. Thus function-architecture co-design technology, and the platform-based design concept, needs to have well-defined interaction capabilities with the appropriate embedded software modelling notations and concepts. This is a key trend for future development of these concepts.

4 SOFTWARE IP IN EMBEDDED SYSTEMS

Carsten Böke*, Carsten Ditze*, H.J. Eickerling°,
Uwe Glässer*, Bernd Kleinjohann°, Franz Rammig*°
and Wolfgang Thronicke°

*Heinz Nixdorf Institut, Paderborn University, Paderborn, Germany
°C-LAB, Cooperation of Paderborn University and Siemens AG, Paderborn, Germany

Abstract

IP-based hardware design became an important topic during the past years. There is an even older tradition of reuse of software components. In this contribution we try to address some key problems of software reuse. First of all, in a bottom-up approach we study some underlying communication techniques used to couple different software components. As the most general solution of this problem is CORBA (Common Object Request Broker Architecture) from our point of view, this technique is discussed briefly. From the bottom we move directly to the top. The most flexible way of handling reusable software IP is to deal with their abstract model. Following this approach, modules can be embedded to various target environments using standard synthesis methods. The problem of protecting property in this approach is not discussed in this chapter. We concentrate on how to combine modules that have been modelled in various languages. At least when using a language coupling approach strict support to define bridging semantics has to be provided. A very useful means for this purpose is given by the ASM (Abstract State Machine) method, which is shortly introduced in the chapter. When software IP has to be used this IP resides on various databases in most cases. We discuss, how design workflows can be defined that allow to access such remote IPs. The exchange format XML plays an important role in this context. Finally we discuss an application example: the application-specific synthesis of real-time operating systems (RTOS) and real-time communication systems (RCOS) from a library of reusable and highly generic software modules.

4.1 INTRODUCTION

Efficiency became one of the key issues in software engineering. It is not surprising that a similar approach as in hardware design was taken: systematic reuse of existing components. This approach is usually called IP-based design in the hardware domain while software people prefer to speak of component based software. The basic concepts, however, are the same ones. One tries to build a library of reusable components and to cover a target design more or less completely by these components. Of course reuse would be no challenge if one could restrict to reuse very simple components, say gates in hardware design or single statements in the software domain. This trivial kind of reuse is common to designers since decades. The challenge is to make complex components reusable, components of the complexity of entire processors or complete database access modules. To make such modules reusable the following prerequisites are essential:

- well defined interfaces of such components
- characterization and description of components
- parameterization of components
- communication platform for component interaction
- frameworks to access and integrate components.

It makes sense to look at the problem of software reuse from the point of view of the inter-component interaction. For this purpose the Object Management Group (OMG) has standardized a platform called CORBA (Common Object Request Broker Architecture). As component based software systems are in most cases built around CORBA (or its equivalent DCOM (Distributed Component Object Model) in the Microsoft world) the structure of interfaces is heavily influenced by CORBA. In Section 4.2 of this chapter some more details of CORBA are discussed.

If components have to be characterized and described, models of these components have to be built. In Section 4.3 we will discuss some approaches for coupling reusable components given in a variety of such models while in Section 4.4 the important problem of semantic foundation of such models is addressed. There are numerous approaches to characterize software components and define similarities. The project SFB501 from Kaiserslautern University seems to be especially promising [Baum98]. Section 4.5 of this chapter concentrates on a special problem: how to build a system that supports intelligent resource management, especially resource reuse. Obviously the probability of a component to be reusable increases by its generality. At the same time its performance decreases. A promising solution of this problem is the use of heavily parameterized components.

Usually components (as IP) are stored in distributed databases, controlled by the owners of the IP. The exchange of the underlying information therefore is a key issue. XML offers promising features. These aspects are discussed in Section 4.6. Finally in Section 4.7 an example of a component based software development system is given. This system makes use of a highly customisable library of reusable operating system components and supports the design of the real-time management and communication part of embedded systems by consequently making use of reuse techniques.

4.2 CORBA: THE BASE TECHNOLOGY FOR INTEGRATION

4.2.1 CORBA History

In 1989 several companies joined to form the OMG to define an open software architecture that provides explicit support for object-oriented design and distributed execution over a network. The result of this effort was the Common Object Request Broker Architecture (CORBA) which defines how distributed software objects can work together independent of client and server operating systems. In 1995 the CORBA 2.0 specification was released which is supported by virtually all CORBA implementations. However, the standardization has advanced since then, reaching 3.0 standard now.

4.2.2 Architecture and Software Development Process

The CORBA standard consists of two parts:

1. GIOP (General Inter-ORB Protocol) which rules the way in which synchronous messages are being exchanged between objects residing in different environments (operating system, network protocol). IIOP (Internet Inter-ORB Protocol) is a derivation of this protocol running atop of TCP/IP. Unlike HTTP, which is a stateless protocol, IIOP allows state data to be preserved. Moreover, the IIOP standard contains a mechanism to uniquely assign identifiers (Interoperable Object References - IORs) to objects residing in the network.

2. IDL (Interface Definition Language) which serves as a basic means to specify object-oriented distributed system in a implementation language independent fashion, i.e. the developer writes his IDL, chooses the preferred target implementation language (C, C++, Java, ADA, SmallTalk etc.) and uses the IDL compiler to turn this specification into server skeletons and client stubs in that implementation language. The translation pro-

cess is regimented by language specific bindings allowing the easy exchange of skeleton/stub code generated by IDL compilers provided by different vendors.

Based on the generated code, the developer still has to integrate the business logic and legacy code but is not longer concerned with identification and communication of the objects, handling transactions and events among objects and securing message exchange within the distributed system.

In order to achieve this high degree automation and convenience, the CORBA architecture is enriched with a set of services (Naming, Event, Transaction, Security Service,...) that help objects interact with each other. Based on these basic services, sets of domain specific building blocks (e.g. CORBA Med for health care or CORBA Finance for the finance sector) can be established.

4.2.3 Components and Multi-tiered Applications

One of the advantages of CORBA is that it provides a path for advancing legacy systems, and other existing code bases, into the cross-platform, Internet-enabled and object-oriented future. This can be achieved by defining appropriate wrappers for the addressed legacy applications and adding the glue logic to the generated skeletons. Having done that, the resulting wrappers can be instantiated as IIOP-enabled, standards-based components. Because CORBA objects can reside on a variety of systems, CORBA is said to enable the proliferation of smart multi-tiered applications redeeming monolithic approaches.

4.2.4 Future Trends

As with all emerging technologies, a true consolidation of the standard can be expected within the next years. The crucial (yet unanswered) question is: what makes a standardization driven middleware technology like CORBA superior to Microsoft's DCOM.

The current situation gives some evidence that DCOM has a main emphasis on desktop-centric computing whereas CORBA is best settled on the server side. There are two major parameters influencing the propagation of CORBA: (a) Within the Java Beans environment there is an explicit support for serialization via IIOP. Moreover, RMI over IIOP has become an integral part of the language. So each development and runtime environment claiming to fully support Java has to be equipped with the IIOP communication protocol. (b) Especially for the server side, the open operating system Linux has gained a growing interest. There are quite a few CORBA implementations running on the Linux platform but there is virtually no support for DCOM.

Not surprisingly, gateways connecting both, the CORBA and DCOM world, have also evolved. Additionally, there are several IDEs (integrated development environments) allowing to define a distributed object system independently of the underlying middleware technology (CORBA, DCOM, RMI) based on UML. The unification of modelling styles for designing distributed object systems based on XMI (XML Metadata Interchange) is also today within the scope of the OMG.

4.3 TYPES OF COUPLING FOR SOFTWARE IP

In this section we assume that software IP is given by high level models. This is an approach of a very high degree of reusability as the reusable modules can be mapped by the usual synthesis steps into the specific target environments. In such an approach models in a variety of modelling languages have to be considered. Here efficient and flexible solutions are required.

One way of composing several specification languages is language coupling. For language coupling the interfaces of subsystems specified in different languages have to be connected. This can be reached by explicitly describing the connections between the subsystems with a corresponding tool or implicitly by naming conventions for identifiers (using the same identifiers in different subsystem specifications). The subsystem models are not changed at all and follow the paradigms of those languages used for their specification. Hence, assertions regarding the entire system are restricted to the subsystem interfaces. The design process works independently on the model parts as depicted in Figure 4.1 (without use of a common model).

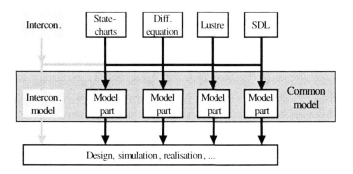

Figure 4.1 Language coupling

Using the concept of language coupling, has usually the following consequences for subsequent steps in the design process. Optimization steps can only optimize the distribution of entire model parts in the overall controller topology or perform local optimizations w.r.t. a single model part. A shift of functionality between model parts specified in different notations cannot be done automatically but has to be done manually by altering the specification. Furthermore, the simulation of coupled model parts causes performance loss due to the communication overhead of (at least) two simulator kernels involved for the corresponding model parts. Commercial tools supporting the coupling of Statecharts ([Hare77], [ILog97]) and block diagrams are already available ([ISI98], [Tanu95]). The scientific community often uses the concepts of language coupling to integrate tools with reasonable effort. An example is the approach described in [Rust98] which is based on Software Circuits.

The goal of a design method should be to support reuse at even higher levels of abstraction with largely automatic generation of realizations from a specification. Language integration assumes that all involved specification paradigms are mapped onto a common internal model (see Figure 4.1).

A common model has to provide constructs that cover all concepts of the specification languages that should be mapped to it. Hence, a common model tends to become very complex and unwieldy resulting in the fact that established design methods cannot be easily transferred to fit with a common model of the required complexity. To overcome these difficulties and transfer at least parts of a design method some restrictions have to be considered during the definition and application of a common model.

Firstly, it should be possible to identify and separate the part of the common integration model generated from a specification (part) in a certain language from the rest of the specification. This results in the requirements for modularity and hierarchy.

Secondly, for such a part in the common model restrictions should be definable in order to allow a transfer of the design methods applicable for the original specification language to this "restricted" common model.

Last but not least, it should be possible to recognize constructs of the original specification languages also in the common model. This feature on the one hand eases the translation into the common model, on the other hand design steps are more easily traceable.

Extended Predicate Transition Nets (Pr/T nets) ([Genr87], [Pete81], [Cher81]) are an example of such a common integration model. We have developed the SEA Environment for specification and animation of extended Pr/T nets. The handling of foreign language constructs is supported by an abstract graphics that can be linked to the underlying Pr/T net ([Klein97], [Klein96]). This graphics can be animated in the desired way when the according net is simulated. By providing appropriate libraries for Statecharts, SDL, block diagrams etc. users can directly use the constructs they are familiar with, without knowledge of the underlying Pr/T net [Klein96].

Currently some other integration approaches are developed. On the basis of UML ([Rati97], [Booc96]) efforts are undertaken to integrate Statecharts, SDL [ITUT92], Message Sequence Charts, and synchronous data flow notations. Hybrid automata [Henz96] combine the specification of event based state transitions (e. g. due to exceeding of intervals) with differential equations. Many specification languages relying on the synchronous paradigm like Lustre, Signal, or Esterel

[Halb93] can be integrated on the basis of a common synchronous automaton model [Maff96]. VHDL and C in the hardware/software co-design domain are often integrated by means of control data flow graphs. However, a commonly suitable solution is not yet known. Here additional research is necessary.

4.4 BRIDGING SEMANTICS: THE ASM APPROACH

4.4.1 Abstract Machine Models

Complex technical systems in virtually all application areas increasingly rely on embedded hardware/software components performing various control operations and supervision tasks. Typical examples are automotive control, production control and extended telecommunication services. In general, one can observe a strong tendency towards distributed solutions running on heterogeneous system platforms which are interconnected through networks. As such embedded systems are characterized by their concurrent and reactive nature; also they often perform operations which are subject to external timing constraints. Additionally, one has to take into account that complex control systems are typically realized as loosely-coupled aggregations of disparate components performing specialized tasks rather than monolithic architectures having a regular structure[Wolf94].

Industrial and scientific engineering of parallel and distributed embedded systems therefore requires integrated specification and design approaches allowing to combine different models of computation and data. Consequently, one obtains hybrid models involving several domain-specific description techniques (dealing with different facets of system behavior). Aiming at a coherent and consistent integration of the resulting model parts, it is a challenging task to find the right abstractions to cope with complexity, diversity and the presence of "dirty" system features (one usually encounters in real life systems).

A systematic description and analysis of embedded system models - regardless whether this is done at an intuitive layer of perception or formally by analytical means - requires a fundamentally changed understanding of the underlying computation models. For such a scenario a notion of computing based on interaction (rather than on "algorithm" and "function" in the sense of traditional computer science) seems to be more adequate [Miln94]:

"We have to admit that computer science - or informatics - is no longer just a matter of prescribing what will happen in the small world of a sequential computer; it must also describe what happens at large in parallel computers, in networks, in hybrid human-machine systems, and even in nature."

Building on the Abstract State Machine (ASM) approach to mathematical modelling of discrete dynamic systems [Boer99], we use here the specific class of multi-agent real-time ASM as a formal basis for meta-modelling. In the context of embedded systems, the objective is to develop a semantic framework providing expressive means for defining a "bridging semantics" allowing a smooth integration of heterogeneously specified system models.

4.4.2 Abstract State Machines

ASM models combine declarative concepts of first-order logic with the abstract operational view of transition systems in a unifying framework for mathematical modelling of discrete dynamic systems. They naturally enable abstract operational interpretations, where the models can be executed on abstract or real machines, providing simulation and test facilities. The scope of applications ranges from distributed control systems [Beie96] over discrete event simulation systems to formal semantics of system modelling languages (like SDL [Glae99] and VHDL [Boer95]). Based on an open system view, both the interoperability of system components as well as interactions with the physical system environment are modelled in terms of relations among interfaces. In that respect ASM models are more like interaction machines in the spirit of [Wegn97]. From a systems engineering point of view, the ASM approach has considerable advantages to cope with the notorious correctness problem that arises in mathematical modelling of non-mathematical reality - namely to construct a model that reflects the system under investigation so closely that the correctness can be established by observation and experimentation. That is, it allows to formalize complex system behavior in terms of executable models using familiar structures and notations from discrete mathematics and computer science in a direct and intuitive way.

4.4.3 The ASM Workbench

The ASM methodology is supported by an advanced tool environment, called the ASM Workbench which has been developed at Paderborn University. The ASM Workbench currently includes a type checker, a simulator, a graphical user interface as well as interfaces to other available tools, e.g. for model checking.

ASM Workbench tools are based on the ASM-SL language (ASM-based Specification Language), which complements the basic language of ASMs - consisting of transition rules specifying behavior - by additional constructs which allow to define the structure of the ASM states, i.e. the data model of the application to be modelled. Such constructs are similar to those found in model-based software specification languages, such as VDM [Jone90].

4.5 REUSE AND WORKFLOWS - TOWARDS INTELLIGENT RESOURCE MANAGEMENT

Using workflow management systems for efficient business processes has become state-of-the-art for controlling, thus guaranteeing efficient working procedures. The term workflow itself inherently describes the steps or activities necessary to perform a certain task. The central idea of business process re-engineering is to find the workflow that yields the optimal (cost and time) solution for a specified problem. Workflow management systems control the repeated executions of workflows which is a most basic form of reuse.

However, the application of repeated working procedures is not limited to pure office applications and document management. Nearly every task in the domain of information processing and engineering involves repeated elements. The problem here is not to reuse a tool but how to reuse a tool.

Workflow management and tool integration actually are specialized solutions for resource management issues. The resources of a certain unit can be characterized by the amount of people with their specific expertise, available soft- and hardware infrastructure and relations to external resources. From this point of view reuse is the key to an efficient resource utilization on the long term.

In this more generalized scenario very different aspects and levels reusability can be identified:

4.5.1 Aspects of reusability for working procedures

Different views are suitable to characterize reusability aspects for workflow management and tool integration environments. Here three of the most important ones are presented: the user-view, the representational view, and the architectural view:

The user view describes on which level of user-interaction re-use is involved. Reuse can be completely restricted to the application development cycle. This means that no concepts of reuse are visible for the end-user of a software product. In this case the application is completely opaque and no data from further program runs can be re-used. Actually most applications support some form of re-use, for instance a clipboard for inputs.

The representational view addresses how reuse is introduced to the user in his environment. Explicit representation comprises the direct access to libraries or repositories of objects and it offers an explicit management.

The architectural view reflects the design of a software system. The structural impact of building a reusable system can be characterized by two main aspects:
- Is the system itself build from reusable parts/components?
- Does the system support reusability for the end-user? Is reuse only an add-on for the user, or has reusability been a guiding principle for the design and specification of the software design?

4.5.2 Levels of reusability for a generalized workflow and integration system

The level of reusability of a workflow and integration system is determined by the granularity of the reused objects. This view mirrors the internal structure of the system.

On the highest abstraction and problem level is the *reuse of workflows by execution*. Consequently a workflow can be reused if the same problem with different parameters occurs which usually implies to execute a sequence of activities with a different set of data. Integrated tools and services are invoked with a minimum of varying parameters. Most of the tasks or steps on this level can be highly automated.

Every workflow consists of a set of resources that are combined by a *workflow definition* specifying the allocation of these resources. *Reuse on definition level* is directed towards the following issues:
- Modification of workflow definitions for similar tasks
- Construction of new workflows by reusing whole or partial workflows for complex tasks.
- Reusing of activities. By this type of reuse the workflow designer selects the appropriate activities for a new workflow from a repository of activities that have already been integrated in other workflow definitions.

The base layer of reuse in a workflow system directly relates to the integration layer. This abstraction always has to provide the infrastructure that allows the environment to control the application processes and to manage the mapping from real data files or objects to the workflow-level.

4.5.3 Knowledge Management Aspects

Capturing complex working procedures through integration and workflow management presents a powerful means of knowledge conservation. Especially in engineering and scientific domains the work heavily relies on a multitude of highly specialized tools and services, most developed for a single purpose. Easy integration and reuse of such tools through workflows reveals the cognitive potential and allows the direct access and facilitated improvement. The workflows actually form a significant part of the intellectual property of an organization.

4.6 DOCUMENT EXCHANGE IN SOFTWARE IP FRAMEWORKS

Reuse has always been a key concept of engineering. Most of the design steps are nowadays carried out in virtual design environments with powerful simulation techniques. In order to reduce the number of employed models generic models are available that can be adjusted in the behavior by a suitable parameterization. On the one hand it is important for IC-vendors for instance to provide models of their devices to engineers on the other hand models and their parameterization constitute a relevant part of the intellectual property of a company.

One aspect of reuse in the area of modelling is the creation of generic models and their parameterization. Reuse and IP issues cannot be realized without using modern internet technologies if more than local solutions are required. XML [XML98] can alleviate some of the reuse problems and offers a viable solution for IP management, too. The use of XML documents as a portable representation for model parameters is flexible and powerul in the perspective of a network-wide distributed environment for reusable objects.

The e*X*tensible *M*arkup *L*anguage is a recent development based on the more complex SGML (Standard Generalized Markup Language) definition which allows the definition of arbitrarily structured mark-up-languages for documents. With XML arbitrary content elements can be marked for further processing. Since XML is a meta-language for defining a mark-up-language, the *Document Type Definition (DTD)* of XML contains a complete description of the mark-up of a document. The representation is de-coupled from the file, and special prescriptions for visualization are kept in a separate XSL file (eXtended Style Language).

4.6.1 P-XML -- The XML parameter set representation

Designing an XML representation has been governed by the following main objective targets:

- *Access of remote entities.* To open a parameter definition to reuse with respect to current internet technologies an established notation for specification should be used. Access to every marked entity of a parameter set should be possible.
- *Clear separation of conceptual blocks in the definition.* The independence of conceptual blocks like "constraints" or "parameters" or "computations" should be preserved in order to minimize side-effects. Constructs like putting a constraint inside a parameter definition as in the old SYBES format are strongly discouraged.
- *Extensibility through modular design.* As new areas for the applications can not always be foreseen and new concepts or advances in information technology have to be respected the representation should be straightforward to augment by new conceptual blocks and constructs.

In the design of P-XML the attribute base can be used, for every entity that should be accessible remotely, to specify the exact location of the remote definition. Additionally every major entity of P-XML may define a unique identifier for referencing it from another document. This is intended for implementing the reuse on the basis of inheritance. All enclosed entities override the inherited sub-entities. P-XML allows only single inheritance to avoid the problem of conflicts.

4.6.2 Reuse and IP management with P-XML

Every entity in a P-XML parameter definition can be referenced remotely. This feature enables reuse of existing parameter set descriptions. If a model is augmented or adapted the accompanying parameterization is derived from the original one. Compound descriptions can reference different parts even from different locations. The mechanism of inheritance offers the most powerful feature for reuse going beyond a simple inclusion and modification. Inheritance incorporates a dependency structure clearly signifying the adjustments made to the parameter set.

Locating parameter sets with URLs allows the realization of web-based client-server architectures for IP management. Accessing IP from a vendor can be coupled with current security and e-commerce mechanisms in available in the WWW. Another perspective is to perform the parameterization of a generic model completely in the domain of the model vendor.

4.7 AN APPLICATION EXAMPLE: BUILDING REAL-TIME COMMUNICATION SYSTEMS FROM SOFTWARE IP

4.7.1 Towards Operating System Synthesis

The overall goal for the design of the DREAMS library operating system [Ditz98] is to support the construction of an execution platform closely tailored to the needs of its application(s). With respect to embedded real-time control, the system must expose a predictable timing behavior and it should run on low-cost hardware that provides "*sufficient*" computation power to meet the deadlines. Therefore it cannot waste any resources like computation time and memory. Consequently, the application must have direct access to the hardware. System services that are not used by a particular application have to be excluded from the execution platform and likewise. On the other hand aspects of reuse, flexibility and portability play also an important role.

We believe that all these conflicting demands can be accomplished by composing the application from a general purpose library of reusable objects that are configured at the source code basis according to the application demands and with respect to the underlying hardware. Rather than developing pure OS-specific mechanisms for enabling customizations, our approach exploits the preprocessor to extend the class concept of C++ by a notion of *structured class skeletons*. By this way, customization is completely independent of the OS, hence it can be applied to arbitrary class structures.

4.7.1.1 Structured class skeletons

Structured skeletons are reusable multi-purpose frameworks based on the ideas of algorithmic skeletons and program families. On a first glance, a skeleton is an ordinary class represented by a name X. It contains some methods and other objects (termed components). In contrast to a class, a skeleton distinguishes three types of components a. Aggregated *mandatory* objects are omnipresent, their type name contribute significantly to the meaning behind the skeleton name. A *customisable component* is similar, but its type is left symbolic during the software development process. The class names of all contained mandatory and customisable components form a vocabulary of characteristic nouns used in an informal *description* of the skeleton name X^1. Therefore, reusability is improved as each skeleton comprises only components whose name contributes to an explanation of the key express of the class name. Conventional classes contain almost always some member objects implementing some *"features"* used by a fairly small number of client classes (e.g. to support accounting). These are meaningless for the description of the class name and hamper reuse as they represent pure overhead for applications that do not use the feature. *Optional components* are introduced in skeletons for this purpose. Their concrete type is also customisable, moreover, the application programmer may decide to exclude these components completely from X. The skeleton name "X", a list of synonyms, the names of mandatory components and those of operations identifying X define a *search index* used to lookup the skeleton in a source code database.

Once a programmer has selected a particular skeleton from a library to compose his solution, he is responsible for resolving the symbolic type names of all customisable and optional components by specifying a known concrete type name for it. The customized type name represents a suitable class or skeleton either available in the library or created by the application programmer.

The last extension to classes are *knowledge attributes*. Knowledge is expressed in a behavior-oriented manner by providing a particular set of methods returning characteristics about the implementation of a class and about performed custonfigurations. As an example for the former, imagine each class like X implements a method *isExclusive()* informing whether it is exclusive or not. If it is, then the implementation of operation(s) provided by X contain critical sections. This knowledge may cause clients of X to protect operations on X against race conditions.

4.7.1.2 Design of the DREAMS library OS

The DREAMS application manufacturing supplement (AMS) is a two-layered library OS. At the bottom layer, Zero-DREAMS is a skeleton for a minimal execution platform supporting only services common to all kind of OSs. In other words,

---

1. Note: A full description of X requires to follow its line of inheritance including the descriptions of all superclasses.

60   VIRTUAL COMPONENTS DESIGN AND REUSE

the mandatory and customisable components provide only services which are always required to prepare an arbitrary application for execution on a native processor. From a top-level view the execution platform is described as an object that has a customisable processor board and an application. The board skeleton in turn consists of a processor, a RAM and different optional device components.

At the second level, a library of further optional classes and skeletons is provided to extend the minimal non-preemptive, single-processor, single-threaded system seamless towards a preemptive distributed multi-threaded application-specific execution platform.

To get an idea on how customization is exploited to balance the tradeoff between flexibility, reusability and performance, Figure 4.2 sketches the design of a "Depot" skeleton for handling the well-known producer/consumer problem. In the semaphore-based solution the depot being shared between producer and consumer is protected by a binary semaphore (Mutex) while over- and underflow situations are handled by two counting semaphores "empty" and "full". According to the class diagram, the semaphores are actually symbols referring to customisable components. Suppose it is known that the producer will never store items into the depot while the consumer is fetching an item and vice versa. In case the depot is implemented based on traditional C++ classes, reusing the source code is overhead-prone as synchronization is not needed in this scenario. In the skeleton approach reuse is free of overhead. The knowledge is exploited by configuring the depot as a specialized *Object* and empty and full as *Counter* objects bypassing the *Resource* and *Semaphore* classes. Consequently, the implementations for acquire() and release() degrade to integer-based operations and the memory required for the depot, empty and full decreases as well. In fact, experience has shown that applications designed based on customized skeletons compare to completely hand-written designs regarding memory and performance.

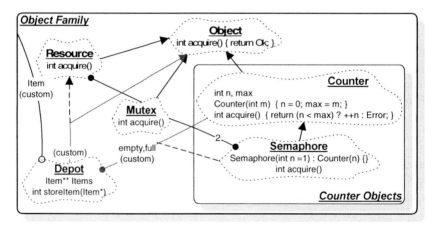

Figure 4.2 "Depot" skeleton for the producer/consumer problem

4.7.2 The TEReCS Project

The design of the real-time communication system as a main part within the whole system is not trivial. Often many iteration-steps in a trial & error manner are needed in order to find a proper solution. Moreover, the communication software is mostly re-written by hand for each new system because a complete re-design due to changes in the application is required. Using *software components* is a feasible approach for supporting *reusability*. The aim of *Component-Software* is to use an automatic synthesis process for the generation of code for a communication system. *Software synthesis* is an alternative approach to real-time system software instead of using kernel-like operating systems. The aggregation of SW components out of a system specification leads to a better utilization of the available time and resources. Opposite to this synthesis technique, operating systems often handle more general applications and therefore are very overhead-prone. Customization of components is a central point for adapting the system software to the demands of the application by exploiting knowledge from the system specification. Component-Software is a suitable approach towards software reuse and portability by using retargetable generators and multi-target component libraries.

In the research project TEReCS[1][Boek99] we focus on the automatic generation of the communication system out of *a multi-target component library* (DREAMS) [Ditz98] for a yet distributed application. Communication services are regarded as components. The whole collection of all needed components for a system represents a *service platform* for data exchange. In this section a short overview over the basic system model and the synthesis process is given.

4.7.2.1 Inter-Component Model-Structure

The interdependencies between all services are modelled in a directed graph of service components. Besides calling dependencies between services, timing dependencies (e.g. an initialization service must be called before an other service) are also modelled. Even hardware dependencies (e.g. a service needs a HW resource just like a device driver needs a device) are included within this graph.

The basic data structure that enables the reuse of software components is the **Universal Service Graph**. Within this graph all interdependencies between all software components are described by special edges; including:

- calling dependencies: a service calls another service,
- activating sequence dependencies: a service must run before/after another service,
- excluding/including features: the use of a special service excludes/includes the use of another service (without calling it).

---

1. Tools for Embedded Real-Time Communication Systems

The *Universal Service Graph* comprises special nodes representing anchors for the synthesis process. These nodes represent *user primitives* that have only a *User Access Point*. These *user primitives* represent no components but an access interface to the *Universal Service Graph*. The totality of all services (components) being developed within our framework are represented in the *Universal Service Graph*. This Graph may consist of many incoherent sub-graphs.

A real system consists not only of software but also of hardware components. We distinguish three different types of hardware components: (1) central processing units (CPUs), (2) devices and (3) media.

Each type of CPU, device or medium is represented by a node in the **Universal Resource Graph**. The nodes are regarded as resources. Edges between resources show that this resources can be connected, that means are compatible. Only one restriction exists: CPUs cannot be connected to media. Edges within this graph are directed showing the possible direction of data flow. Each type of CPU, device or medium that will be used in a system must have a representation as a node in the *Universal Resource Graph*.

The *Universal Service Graph* and the *Universal Resource Graph* are the basic data structures which describes all components out of which a system can be build. The combination of the *Universal Service Graph* and the *Universal Resource Graph* leads to the **Universal Resource Service Graph**.

4.7.2.2 System Specification

In this paragraph the way how a desired system is specified is presented. This specification serves as input for the synthesis process generating the communication system code. It has to be given by a user. The specification of a system to be synthesized out of components is described also by two graphs: The **Resource Graph** and the **Communication Graph**.

The **Resource Graph** represents the hardware components and their connections available in the system. Each resource (node) of this graph represents a concrete instantiation of a *resource type* existing in the *Universal Resource Graph*. Each edge in this graph is only valid if an edge between corresponding resource types in the *Universal Resource Graph* exists. In other words the aggregation of the *Resource Graph* - where all nodes of the same type are combined - must result into a sub-graph of the *Universal Resource Graph*.

The **Communication Graph** describes the distribution and the communication behavior of the application. Each node of the graph represents a process of the application. An attributed directed edge within the graph shows a possible *communication connection* between these processes. The edge is attributed by two user primitives (the sending and receiving one), the amount (and eventually the type and layout) of the data and the period or maximal allowed latency for the communication.

4.7.2.3 Synthesis Process

After presenting the basic data structures used for the description of the SW-components the synthesis process is shortly introduced (see Figure 4.3). We assume that the distribution and mapping of the threads onto the CPUs of a target system is already done. As well we assume that the resources of the target system and their connections are also pre-defined. As input for the synthesis process serves the *Communication Graph* and the *Resource Graph* specified by the user and the *Universal Resource Service Graph* as a description of the library of all available components.

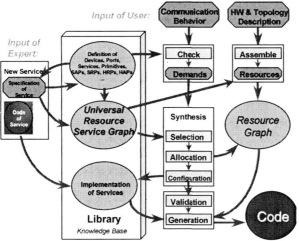

Figure 4.3 Synthesis process for component-based software

The first step (**Graph-Composition**) of the process combines all this three graphs to one hybrid ***Resource Service Graph*** for the required system. This is done in the following manner:

- All resources in the *Resource Graph* are connected to their corresponding types in *the Universal Resource (Service) Graph*.
- Each user primitive in the *Communication Graph* is assigned to its corresponding *user primitive type* in the *Universal (Resource) Service Graph*.
- All paths in this graph ending in *user primitive types* that are not used are deleted.
- All paths in this graph with services leading to *resource types* that are not connected to *resource instances* of the *Resource Graph* are also deleted.

This resulting ***Resource Service Graph*** consists only of probably required services and resources for the demanded system. The second step of the synthesis process contains the main algorithm for finding the really needed services and resources.

The main idea is that this step works like hardware synthesis. So it is divided into three sub-steps.

The first sub-step (**Selecting & Allocating**) determines the really needed services and resources. This is done by finding an optimal path in the graph from the sending user primitive to the receiving user primitive of each edge of the *Communication Graph*. User primitives lead to services, services lead to other services or device types, device types lead to concrete devices on this CPU, devices lead to media, media again to devices, devices to device types, device types to services and services again to user primitives. In this way, by searching a path through this graph, the routing-problem is solved too. Because of possibly existing alternatives in the *Universal Service Graph* and in the *Resource Graph* more than one path can be found. The optimal path is chosen by using a cost function. By finding a path the selected services can be allocated onto the corresponding CPUs. Furthermore, it is known which services and resources are used by which communication connection. This way, the resource loads of all devices and media (and even CPUs, if we know the generated load by each service) can be determined.

The second sub-step (**Generating & Configuring**) generates data which are needed for the services (e.g. a routing table for a routing service) and assigns free parameters of services to concrete values (e.g. baud rate used for medium).

The third and last sub-step (**Collecting & Compiling**) builds up the code for each CPU. This is accomplished by generating Makefiles which compile the needed services and generated data into a library for each CPU.

Tools working on these data structures and using the above presented algorithms are able to find and create the code needed for a real-time communication system for a distributed application for embedded micro-controllers, if all needed components are available. If not, the tools can give hints on which components are missing in order to find a solution: This may be indicated by missing edges in the graphs. Furthermore, if a solution is found, the tools can calculate the resource loads. By this means it is possible to detect bottlenecks in the system.

5 ARDID: A TOOL AND A MODEL FOR THE QUALITY ANALYSIS OF VHDL BASED DESIGNS

Yago Torroja, Felipe Machado, Fernando Casado, Eduardo de la Torre, Teresa Riesgo and Javier Uceda

Universidad Politécnica de Madrid
Madrid, Spain

5.1 QUALITY ATTRIBUTES OF VHDL DESIGNS

The quality of a product, i.e. a digital design, is a fuzzy concept that can be considered from different points of view. Usually, the most accepted definition of quality is the fulfilment of explicit or implicit requirements. Although there are several definitions of quality ([Cros79] [Demi86] [ISO87]), "quality can not be well defined, but it can, and should be modelled" [Jozw96].

In a broad sense, the final quality of a product can be understood as the result of the quality of the design process and the quality of the product itself. The quality of the design process is influenced by several factors: management, design skills, available tools, good documentation, design methodology, etc. In order to assess the quality of the design process it is not only necessary to analyse the process itself but also to measure the quality of the product at every moment.

In the case of VHDL based designs, the quality of the final product can be analysed considering the VHDL descriptions. As it has been commented, the features to consider depend on the quality requirements of the design. Anyhow, there is a common set of features normally considered ([TOMI96] [Sina94] [Keat98a]). Instead of giving a precise definition of these features, a more "for use" definition applied to VHDL descriptions can be summarised as follows:

- **Maintainability**: it is an indicator of the effort needed to modify or correct a design. Maintainability is mainly related with readability, design organization and design tool capabilities.
- **Readability**: it reflects the ease of a description to be read and understood by a person. It is related with aspects like complexity, name conventions or profusion of comments.
- **Complexity**: it reflects how difficult it is, or has been, the development or interpretation of a description. This feature is related with aspects like code size, nesting level, degree of modularization, etc.
- **Portability**: it is a view of the ease of a design to be used in different environments. It can be considered from different points of view, portability between tools, between target technologies, between users, applications, etc.
- **Reusability**: it is an indicator of the ease of a design to be used as a part of a bigger design, or to be adapted to a new application. It is related with aspects like portability, maintainability, and degree of parameterization or ease of integration into the design flow.
- **Simulation performance**: it reflects the efficiency of the simulation process for a design. Code complexity, modularization degree, or number and type of data objects are factors that directly affect the simulation performance.
- **Compliance with guidelines**: this feature reflects the degree with which certain rules and guidelines have been followed during the development. These guidelines can affect, among others, name convention, design style, code complexity or any other feature of the design.
- **Synthesis efficiency**: it reflects the quality of the hardware obtained when the design is synthesized. It can be considered from the point of view of design performance (in terms of area, power consumption or delays), or from the reliability of the resulting circuit.
- **Testability and verifiability**: testability is related with the ease for a design to obtain a set of patterns for manufacturing test. Verifiability reflects the ease to develop a good test bench to verify the design functionality.
- **Reliability of the hardware description style**: there are descriptions that may lead to error-prone designs when synthesized, resulting on a circuit with a different behavior when compared to simulation. Reliability considers the presence of this kind of descriptions in the model.

In order to assess these features, it is necessary to identify some attributes of the descriptions that can be quantitatively or qualitatively measured. These attributes can be subsequently divided in sub-attributes that are easier to measure or assess. Finally, the set of measures obtained from these sub-attributes will be the input of a function that globally assesses the considered feature.

Several VHDL quality analysis methods can be found in the literature. Some of them are static analysis methods (mainly those imported from software engineering) ([Mast96] [Obri97] [Bare96]), while other use dynamic techniques (specially those dealing with functional validation) [Ries97]. Most of the quality analysis methods are based on checking the compliance of VHDL descriptions with respect to certain VHDL coding guidelines.

On the contrary, most of the quality checkers included in ARDID (and described below) are mainly based on the analysis of a simplified version of the hardware that will be synthesised from the descriptions, as it will be seen in Section 5.3.

Our efforts have been mainly focused in measuring the quality of the hardware that will be obtained after synthesis, trying to develop some metrics and procedures that lead the designer to a successful circuit. The final goal is for the designer to have a continuous quality control from the first stages of the design. This continuous control may avoid later loops during the design flow and produces a better overall design quality, shortening design cycle. Mainly, three aspects of the design quality have been considered:

- Reliability of the hardware description style.
- Quality of the test-benches used to validate the design.
- Reusability, readability and maintainability of VHDL descriptions.

All these aspects are integrated in a tool that makes a continuous verification of the design quality during the whole design process.

5.2 ARDID: A VHDL QUALITY ANALYSIS TOOL

ARDID is a graphical front-end environment specially designed to work with VHDL designs. The main objective of the tool is to provide the designer with methods to increase the quality of the final design and to have automatic methods to help in the review during the architectural and logic design stages of a VHDL design process.

The tool includes four main functions: a source code Version Control System, a VHDL Library Manager, a VHDL Design Quality Toolkit, that includes several quality checkers, and a VHDL Validation Quality Tool, that analyses the quality of the test-benches used during the development. The tool also integrates other capabilities that allow the user to call external programs (simulation, synthesis) as part of the environment. In the following paragraphs, a brief description of each part is given.

5.2.1 The Version Control System (VCS)

In our experience, there is a lack of culture in VHDL designers to use any version control system. This is normally due to the lack of integration of these tools under the same design environment they use. For this reason, a source code Version Control System has been integrated under ARDID. The Version Control System is a graphical shell on top of GNU's Revision Control System (RCS) [Tich91]. The shell allows the user to graphically see which files are under control, which are editable, only readable, easily check in and out files, see differences, and almost all functionality that RCS has.

5.2.2 The VHDL Library Manager (VLM)

In order to easily apply the quality checkers and manage the design, ARDID includes a graphical library manager with the most typical functions normally included in this kind of tools. The user can create, delete or link libraries, compile source code into the library, edit source code, update compiled versions and obtain scripts to automatically recompile the libraries. The Library Manager is a graphical shell on top of LEDA's LVS System [LEDA98].

5.2.3 The VHDL Design Quality Tool Kit (QTK)

The VHDL Design Quality Tool Kit contains the checkers that are used to measure the quality of the design. These checkers have two main goals. The first one is to detect design methods or VHDL constructs that are likely to produce problems in latter phases of the design. These are hardware-oriented checkers. The second goal is to improve the maintainability, readability and portability of the VHDL code. These are coding-oriented checkers. If all this were done without automated tools, it would imply a detailed and tedious analysis of the VHDL code. The checkers will greatly help in this task, detecting this kind of problems and helping to obtain an overall view of the design quality in an easy way. Quality is measured based on its accordance with a set of rules, stated for expert designers, that will avoid unpleasant surprises at the end of the design process.

Reusability is another important objective of every design. The high cost associated with a design is forcing customers and design centres to invest a lot of effort on making designs reusable. This implies the design should satisfy certain requirements in order to do the design industrially reusable [Torr97]. The use of the VHDL Design Quality Tool Kit will help the designer analysing his/her designs in order to fulfil some of these aspects.

The use of these checkers is oriented towards designers, reviewers or project managers and final IP users. For designers, the checkers will help them to produce better VHDL descriptions and to reduce the number of iterations in the design process. For project managers or reviewers, the checkers, and the compact way the information is reported, will help them to assure that the design team is working properly and according to the organisation design methodology. For the final users, the checkers will help them to analyse the quality of the macro-cells they are using without the need of a deep knowledge of their functionality.

The results of the checkers are presented to the user in different ways. First, it is possible to obtain a hierarchical text description of the results. Through the use of hyperlinks, the user can access to a more detailed information of the result, that can be back-annotated on the VHDL source code. Additionally, the user can obtain a graphical view of the design and the results of the checkers, having a quick method to detect potential problems.

In the current release, the VHDL Design Quality Tool Kit provides the following checkers to analyse the VHDL description:

HARDWARE-ORIENTED CHECKERS

Clock and Reset Analysis: Clock and reset schemes are one of the most critical issues in a design. For designers, it is important to know not only which signals are registered by a clock, or initialised by a reset signal, but also whether the clock or reset may cause problems due to glitches, set-up or hold time violations, etc.

For every clock described in the design this checker points out:

- Whether the clock comes from combinatorial or sequential logic, or from a port
- Signals and variables triggered (or enabled) by the clock signal
- Whether the clock drives combinatorial logic, flip-flop data inputs, etc.

For every reset described in the design this checker points out:

- Whether the reset comes from combinatorial or sequential logic, or from a port
- Signals and variables initialised by the reset signal
- Signals and variables registered but not initialised
- Whether the reset is source of combinatorial logic, flip-flop data inputs, etc.

Clock Domain Analysis: Different clock domains are a source of problems in designs with several clocks. It is important for designers to know which signals are in the frontiers between clock domains, in order to pay special attention to their behaviour. For every registered signal in the design, the checker analyses if the data input of a memory element registered by one clock comes from a memory element registered by another clock, and points out the signals and the clocks implied in the clock frontiers.

Registered Objects Analysis: In the design process, it is important to know which signals are going to be registered by a flip-flop or latch. This is normally known after synthesis, but the synthesis process usually takes too much time to use it as a check method. This checker quickly detects what signals or variables will be registered with a flip-flop or a latch.

Registered Outputs Analysis: When a design is being synthesised in a module or block basis, timing problems may arise when putting all modules together. These problems are minimised if the output signals of a module come directly from a flip-flop. This checker analyses the outputs of the modules and returns:

- Outputs coming directly from a flip-flop.
- Outputs coming directly from a latch

- Outputs that are combinatorial function of the module inputs
- Outputs that are combinatorial function of other internally registered signals

Combinatorial Feedback Analysis: Usually, designers do not insert combinatorial loops when they describe a module. But this situation is more difficult to detect when the combinatorial loop is due to the connection of several modules in a design hierarchy. Most of the synthesis tools detect this kind of problems when they deal with the whole circuit. But again, the time spent in the synthesis could be prohibitive to apply this check regularly.

Hardware Signal Usage: This analysis reports signals, ports and variables that, although declared and used in the VHDL description, will not generate any hardware when the design is synthesised. This checker eases the detection of hidden errors and the understanding of the functionality of the design.

High Impedance Signal Analysis: In some ASIC design methodologies or technologies, the use of tri-state signals is not allowed or recommended. This analysis will highlight signals that are going to be synthesised as tri-state buffers.

VHDL CODING-ORIENTED CHECKERS:

Sensitivity List Analysis: When developing VHDL code, sometimes it is difficult for designers to be aware of errors or omissions in sensitivity lists. This may cause incorrect simulation results that are only discovered after synthesis, delaying the project development. The Sensitivity List Analysis points out those erroneous or "to pay attention" processes, providing the following results, depending on the signals read in the process:
- There is no sensitivity list
- There are only clock and reset in the sensitivity list
- There are signals read in the process that are not in its sensitivity list
- Reset or clock signals do not appear in the sensitivity list of a sequential process

Architectural Description Style: the description style has influence on code maintainability and synthesis results. On the other hand, behaviour embedded in structural descriptions makes the code cumbersome. This analysis will help designers by providing them with a classification of the unit checked according to the following possibilities:
- Description with component instantiations only
- Description with component instantiations and simple signal assignments
- Mixed descriptions with components and behavioural statements
- Behavioural descriptions
- Data-flow descriptions

Hard-Coded Integer Values: Code reusability advises hard-coded values in the description to be substituted by constants or generics. In this way, modification of the module characteristics may be carried out easily. With this checker, integers hard coded in the VHDL descriptions are detected and highlighted. A "clever" algorithm is used in order to report not all values, but only those interesting for the designer (e.g. the tool does not report lower limit in range constraints, like in Bus-Witdh downto 0, or values in increment/decrement expressions of counters, like in Count <= Count + 1).

Initial Value Analysis: Initial values are not taken into account by synthesis tools, but they are considered by simulation tools This may cause differences between the simulated and the synthesised model, and produce new loops in the design flow. This checker highlights the ports, signals and variables that receive an initial value in the code, helping designers to detect potential problems.

Nested if/case statements: When the designs become complex, the sequential logic grows, and a great number of logic conditions appear in the code, making it difficult to understand. This simple checker analyse the number of logic statements that are nested one inside the other, and reports the maximum number of nested statements of each unit to give an idea of its complexity.

Object usage: While coding VHDL descriptions, designers often declare objects that are not used anywhere. This aspect worsens not only code maintainability but also the detection of hidden errors. The Object Usage Analysis reports unused ports, signals, variables, constants or generics.

There are other checkers that are under development; these checkers are mainly oriented to detect redundant memory elements and parts of the circuit that could be difficult to test. Besides, a fast area estimation tool is under test. The objective of this tool is not to obtain an accurate estimation of the gates used in a circuit, but to reliably compare two VHDL descriptions in terms of area. Nevertheless, the tool architecture is open, and new checkers may be developed with relatively low effort.

As it has been mentioned, the graphical user interface shows the results of the checkers in a compact tabular form with hyperlinks. This allows design managers to quickly evaluate and document the quality of a VHDL description or library. The information, warnings or source of problems provided by the checkers is directly highlighted on the VHDL source code. It is also possible, for a quicker view, to show the global results of the checkers as a coloured map in the hierarchy of the design. Figure 5.1 shows the outlook of the QTK.

5.2.4 The VHDL Validation Quality Tool (VTK)

The goal of simulation reviews is not only to check that the functionality is in accordance with the specifications, but also to analyse the quality of the test benches used in the simulations. Simulations that do not exercise all the lines of the VHDL code are normally rejected. This quality criterion may not be enough to guarantee the high quality needed in complex designs, as shown in [Ries97a]. The use of the VHDL Validation Quality Tool (VTK) allows not only to check if all lines of VHDL code are exercised, but also to check if the effects of the simulations may be observed at the outputs of the blocks that are being analysed.

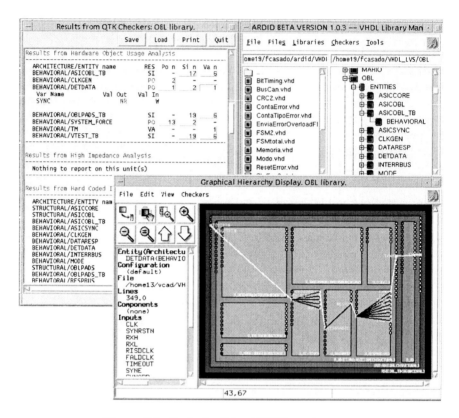

Figure 5.1 Outlook of the ARDID environment

For the designer, VTK helps in the development of quality test benches mainly during the architectural design phase. Anyhow, it allows the designer to detect typical coding errors at any stage. This aspect is specially important due to the great amount of modifications a VHDL model may pass through (even if it is functionally correct).

The VHDL Validation Quality Tool is based on an error model and fault simulation technique [Ries96]. The error model considers typical errors a designer could make when describing a circuit in VHDL. The tool obtains faulty descriptions of the circuit that are compared, during simulation, against the error-free description. Then, an error coverage measure, similar in concept to the fault coverage for manufacturing test, is obtained. The VTK back-annotates directly on the source code those errors that are not detected by the given test bench. For the moment, a commercial VHDL simulator is used to perform the quality analysis.

The tool provides configuration parameters that allow the user to control the type of errors to be considered, the unit where to insert the errors and observe output values, the percentage of errors to simulate, and the simulation stop conditions.

5.3 INSIDE ARDID: SIMPLIFIED HARDWARE MODEL

In order to process the VHDL descriptions, all these tools have been implemented on top of LEDA VHDL System (LVS), that provide access (through an application procedural interface) to the VHDL intermediate format (VIF). Most of the checkers described in Section 5.2.3 provide information about the hardware that will be obtained from the VHDL descriptions. Since the VHDL intermediate format has no information about the resulting hardware, a simplified hardware model with the basic outlines of the hardware obtained after synthesis has been developed. This simplified hardware model (SHM) can also be accessed through an application procedural interface, easing the elaboration of analysis procedures and checkers related to hardware. The main goal when SHM was developed was to be accurate enough to be helpful, while maintaining a simple representation and fast processing time.

In the model, memory elements are explicitly represented, with their inputs (clock, asynchronous initialisation and data) and outputs connected to other memory elements or ports of the circuit. On the other hand, combinatorial logic is not detailed, considering only dependencies between signals, and only differentiating it from direct and inverted connections. This makes a representation similar to the one shown in Figure 5.2, where the hardware is represented as a network in which the internal nodes are memory elements and the boundary nodes are ports.

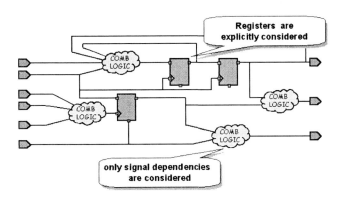

Figure 5.2 Simplified hardware model

This kind of representation highlights only the most relevant features of the hardware scheme, being in some way similar to a graphical RTL representation. The semantic followed to obtain this model is mainly the one proposed in the

74   VIRTUAL COMPONENTS DESIGN AND REUSE

IEEE1076.6 standard for VHDL RTL synthesis [IEEE98]. The simplifications adopted to obtain this model make the processing very fast and let the designer use it from the very first stages of the design process. In some cases, these simplifications may induce differences between the SHM and the real hardware that should be taken into account. Nevertheless, this model should be understood as a method to ease and accelerate the assessment of the hardware, and if the way to obtain the SHM is known, these differences can be easily detected.

In the following, the main simplifications adopted to obtain the SHM from the VHDL descriptions will be explained, exposing the rationale of these simplifications and the cases where they may lead to differences with respect to the real hardware.

5.3.1 Signals and variables considerations

Every signal of the circuit (whether it is a VHDL signal or variable) is considered as a single element, not taking into account if the signal is a bit, a vector or a composite signal, and therefore, not considering indexes or slices. The main advantages of this approach are the reduction of processing time and the possibility to use the model with non-elaborated designs, where the length of vectors or composite signals depend on generics. Obviously, this consideration may lead to erroneous results in the case of signals that comprise several bits, as each bit can be considered individually.

```
. . .
    signal s_array : bit_vector (0 to 7);
. . .
    P: Process( Clk, Resn )
    begin
       if Resn = '0' then
          s_array (1) <= '0';
       elsif Clk'event and Clk='1' then
          s_array <= some_function;
       . . .
```

Table 5.1 Different usage for the elements of an array

As an example, consider the Table 5.1. Although it is not a good VHDL description style, the designer could use each element of a vector in a different way, modelling different hardware for each bit. In this case, the designer has initialised only one bit of the vector.

As SHM will only consider the signal as a single object, an analysis of the initialisation of the signal would report that the signal is initialised, what is not true (at least for some bits). In this case, the user must be aware that the SHM may differ from the synthesised design. Anyhow, in most cases this kind of signals is used as buses with similar connections and usage.

This simplification has other important consequences that will be more clearly exposed in the following paragraphs.

5.3.2 Loop considerations

As a consequence of considering signals as single objects, the SHM does not replicate the hardware in loop statements. Only dependencies between signals are considered, not taking indexes into account. Again, this may lead to erroneous results in some cases, specially in combinatorial logic.

Consider the example in Table 5.2. After synthesis, signal *RD* forms as many registers as the number of its bits, connected as a shift register. None of the individual bits depends on itself, but on other bits of the register or on the input (S_in). But when generating the SHM, indexes are not processed. Therefore, signal RD appears as forming a single register, influenced by itself and S_in, and influencing S_out. This result is not really incorrect, as we can consider that the data of a shift register depends on its own value.

```
...
   elsif clk'event AND clk = '1' then
      S_out <= RD(0);
      for i in 0 to NUM-1 loop
         RD(i) <= RD (i+1);
      end loop;
      RD(NUM) <= S_in;
...
```

Table 5.2 Loop statement

Nevertheless, if we consider a similar example but with combinatorial logic (outside of a clock expression), the SHM would reflect a combinatorial loop affecting the signal RD (that will depend on itself), but actually, there is not such a problem (the description will be synthesised as a single wire). This result should be taken into account in order to assess correctly the information of the SHM.

5.3.3 Generate statement considerations

Conditions and iterations in generate statements are not considered, hardware included in those statements is implemented once and only once. The main reason

76   VIRTUAL COMPONENTS DESIGN AND REUSE

of this simplification is that, in most of the cases, elaboration is needed to know the values of the conditions or iterations of the generate statements.

The results of the SHM have to be considered bearing in mind that, in iterative generate statements, the synthesised circuit may have some replicated hardware and, in conditional generate statements, some hardware may or may not be synthesised (see Table 5.3).

```
...
-- Conditional generate
GEN1: if i=param generate
   -- some optional logic
   ...
end generate;

-- Generate with iteration
GEN2: for i in 0 to param generate
   -- some iterated logic
   ...
end generate;
...
```

Table 5.3 Generate statements

In our opinion, this view, where hardware is represented without taking into account if it is replicated or may not appear, is closer to the designers' view, and this does not represent a real drawback.

5.3.4 Expressions considerations

All expressions in the VHDL description are considered different and are not evaluated. This simplification affects the way latches are detected. As an example, consider an if statement that does not have an else condition (see Table 5.4). To detect if a signal (i.e. *L_sig*) will infer a latch it is necessary to see if it is assigned in all possible execution paths.

In the example, the condition expressions are mutually exclusive, so the signal *L_sig* is assigned whatever value *A_bit* has. But, since expressions are not evaluated during the generation of the SHM, it will reflect that a latch has been inferred, while the synthesis results will not (seeFigure 5.3). If an *else* were used instead, the results of SHM would have been identical to synthesised hardware.

Nevertheless, in our experience, and depending on the expressions and synthesis tool, the results in this case could be correct (some synthesis tools do not detect the simplification). On the other hand, this code would be better described removing the last condition.

ARDID: QUALITY ANALYSIS OF VHDL BASED DESIGNS

```
...
   signal A  :  bit;
...
   if A = '0' then
      L <= '0'
   elsif A = '1' then
      L <= B;
   end if;
...
```

Table 5.4 If statement covering all options without else

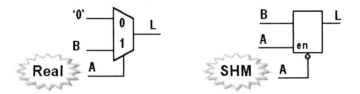

Figure 5.3 Differences between synthesis and SHM for Table 5.4

5.4 EXPERIMENTAL RESULTS

5.4.1 Results from the Design Quality Tool Kit

The VHDL Design Quality Tool Kit has been used to analyse the quality of several industrial and academic circuits developed in VHDL at UPM labs. In order to have a global overview of the quality of a design, a preliminary global qualification has been obtained for every design, taking into account the results of all checkers. Values (see Table 5.5) have been obtained to evaluate maintainability (Mt) and reliability (Rl) of a design (or design library). These values have been obtained using the following weighted sum of the results of each checker:

$$C_T = max(100 - \sum w_i q_i, 0)$$

Where q_i is the number of warnings or errors detected by the checker i, and w_i is a weighting factor for this checker. These factors have been set experimentally according with our design experience, and considering our subjective view of what should be considered a high quality design (e.g. errors in the sensitivity list are strongly penalised, while hard coded values are slightly penalised).

Design Type	LOC[a]	Gates[a]	Mt[a]	Rl[a]
Aerospace ASIC	6800	7000	VH	VH
Industrial ASIC	9200	25000	H	VH
Aerospace ASIC	10200	4000	H	VH
Industrial FPGA	1200	900	H	VH
Academic FPGA	1700	10000	M	VH
Industrial Module Library	8000	200-3000	H	M
Academic Module Library	6100	500-4000	L	M
Industrial FPGA	4000	8000	H	VL
Academic Module Library	8500	3000-8000	M	VL

a LOC: Lines Of Code; Mt: Maintainability; Rl: Reliability;
VH: Very High; H: High; M: Medium; L: Low; VL: Very Low

Table 5.5 Results from QTK for some designs

It is interesting to observe that designs with higher scores correspond to industrial designs with better design methodologies (ESA, PRENDA) ([Sina94] [PREN96]). It should also be noted that some of the designs with very low values did not work properly when the prototypes were tested. In these cases, the source of problems was quickly detected with the help of ARDID.

5.4.2 Results from the Validation Quality Tool

The VHDL Validation Quality Tool has been tested with several VHDL designs, ranging from very simple circuits (finite-state-machines, shift registers, ALUs, etc.) to industrial designs (ASICs implying thousands of gates).

For the moment, the main drawback in the VTK is the simulation time. As no specific error simulator has been developed, a commercial one is used. One way to solve this problem is to select a percentage of errors to simulate in the first tries. The results obtained with this percentage can be extrapolated to the whole number according to the results of our experiments. Finally, if the simulation time allows it, 100% of errors can be simulated. Better ways of inserting errors that could enhance simulation times are under study. In Table 5.6, experimental results are detailed (times are for a Sun Sparc 20).

ARDID: QUALITY ANALYSIS OF VHDL BASED DESIGNS 79

As it can be seen (and could have been extrapolated from fault simulation techniques), the simulation time is greatly influenced by the number of errors, the simulation cycles, and error coverage finally achieved.

Design	LOC[a]	ExS[a]	SmC[a]	Err[a]	Cov[a]	PrT[a]	SmT[a]
SHR2	31	9	18	42	100	1	0
FSM4	70	39	30	122	90	9	1
TRANS	60	29	56	88	91	5	0
RECEV	65	32	62	101	87	7	1
SHREG	41	14	68	58	100	2	0
ALU	196	56	78	109	100	8	1
ALU16	196	84	78	109	100	9	1
MULTIS	81	40	90	168	97	21	2
ASIC53	399	112	12.160	695	100	406	752
ASIC52	310	128	12.160	829	47	460	4.637
INTERR	3286	1046	72.386	1290	100	1.132	14.675
CONT16	43	15	131.084	60	98	3	526

a LOC: Code Lines; ExS: VHDL Statements; SmC: Sim. Cycles;
Err: Errors Inserted; Cov: Error Coverage (%);
PrT: Preprocesing Time (s); SmT: Simulation Time (s)

Table 5.6 Results from VTK for some designs

5.5 CONCLUSIONS

A general view of a tool, called ARDID, for the quality analysis of VHDL based designs has been presented. As it has been shown, this tool is aimed to improve the quality of the designs from the early stages of the design process. This may result in a reduction of design time and higher reliability.

The tool is not only oriented towards designers, but also, and very specially, to project managers that need to control the quality of the design without having a deep knowledge of the circuitry and the descriptions that are being developed. This is not only important when designing complex circuits, but also, when designing with third party modules.

The tool has been tested with several industrial and academic designs with good results in terms of time, ease of use, and accuracy of the results. Although preliminary experiences are promising, at the moment, the whole tool has not been used in an on-going project from the beginning, nor a comparative method has been applied to measure the save of design time or quality improvement due to the use of the tool. Anyhow, we think this tool may greatly help to improve the global quality of a design and to reduce the development time.

6
A VHDL ANALYSIS ENVIRONMENT FOR DESIGN REUSE

Claudio Costi and D. Michael Miller

University of Victoria
Department of Computer Science
Victoria B.C., Canada

6.1 INTRODUCTION

Due to productivity pressures, the IC industry is showing increasing interest in the reuse of components. Three different models for managing and developing components to be reused are of interest [Keat98].

In the *As-is reuse* model, a component is designed for a specific project and then it is put in a common repository. No engineering overhead is necessary since no design for reuse properties are considered. Unfortunately, designers are on their own when they need to integrate the same component in a new design. In fact, in this case, designers need to review and understand all the details of the component since no support is generally obtained from the original designers.

In a quite similar approach, called *rework-based reuse*, the component is again designed for a specific project but, after completion, a dedicated team re-engineers the design to make it reusable. The dedicated team faces the problem of deeply understanding the original design and re-structuring it with design for reuse characteristics.

In the third model, called *IP-based reuse*, a dedicated design team explicitly implements a component with design for reuse in mind. Other teams may reuse the component in any design and, generally, a technical department is in charge of supporting the new users of the component. The IP-based reuse model is perhaps the most effective for design reuse but it is not often used in IC design centres because of the additional costs of designing for reuse, costs which can not be necessarily attributed to one particular design project.

The second and third models have only recently been considered in industry and most teams in design centres are still struggling to reuse legacy designs in an As-is reuse approach. Legacy designs are components implemented with no design for reuse in mind but which have valuable IP content.

6.2 REUSE OF SOFT COMPONENTS

The term *soft component* (or soft model) [VSIA96] is used to indicate a component which is available in the form of synthesizable Hardware Description Language (HDL) code. VHDL (VHSIC - Very High Speed Integrated Circuit - Hardware Description Language) and Verilog are the two most commonly used hardware description languages. Such technology independent description allows designers to flexibly reuse components in IC designs. Many legacy designs are in the form of soft components.

Reuse of soft components has been approached by the IC design community and university researchers in a variety of ways. In [Keat98], RTL coding guidelines for the design of components for easy reuse are presented. These guidelines address the development of new components rather than the reuse of legacy designs.

Organization, maintenance, and retrieval of components from a library database have been the subjects of various research studies. In [Koeg98], a formal specification of a library of reusable components is presented with each component represented by a behavioral description. Given a formal behavioral specification, it is possible to automatically locate a component with the same, an extended or a similar specification.

Since VHDL soft components are among the more common reusable descriptions, research has been done on the management of a library of VHDL modules. In the tool set presented in [Olco98a], library, package and configuration concepts, which are specific to the VHDL language, are used to organize the repository of elements and in presenting them to the user. Versioning, team work management, internet and intranet retrieval capabilities are supported. Similarly in [Prei95] a reuse scenario is presented in which parametric VHDL descriptions are complemented by scripts to invoke different tools. The scripts provide verification data for timing and functionality. In [Reut99] classification categories for objects in the library are defined to help the designer in selecting the required component.

While the above noted research projects assume that specific knowledge is available and attached to an existing component description, other studies use various techniques to directly analyze VHDL descriptions and automatically extract some code information. In [Clar98] and [Iwai96] the use of program slicing techniques to extract subsets of VHDL code, which can be analyzed by designers for debugging, maintenance, functional testing and design reuse, is described. Program slicing is a method of extracting a subset of code from a given program using a specific slicing criterion. For example, for a selected signal assignment line in a VHDL source file, it is possible to extract a slice which consists of all the code lines which directly or indirectly affect the chosen assignment line.

A more semantic related approach to VHDL code analysis is presented in [Bare96]. The goal is to identify properties of the final circuit from the synthesis and testability points of view. The presented analyses are able to identify signals which are implemented as memory elements in the final circuit and signals that can be used to propagate test patterns through a portion of the design.

To the best of our knowledge, the only previous research work which explicitly aims to help designers in understanding VHDL code is the one presented by Gunther Lehman *et al.* ([Lehm96a] [Lehm96b]). In their work, a VHDL reverse engineering tool-set called VYPER! is presented. In VYPER!, five different interfaces graphically represent information about the VHDL code, including: (i) control flow of concurrent statements; (ii) control structure of sequential descriptions, and (iii) the module hierarchy. By means of hyperlinks, the designer may easily investigate the VHDL code and search for information which is necessary to reuse or modify the component under analysis. The goal is to help designers in analyzing and understanding the code itself and there is no attempt to identify or assist designers in understanding what the VHDL code describes.

6.3 VALET PROJECT

Motivations

As mentioned above, design for reuse methods, by which components are specified and developed with reuse in mind, were not and, often are still not, implemented in many IC design centres. Therefore, a large number of legacy designs exist with valuable knowledge in them but with little if any information regarding their potential reusability. Often only partial or out of date documentation is available and designers must recapture the original design intent directly from the component description.

Even in cases when a good functional specification and test exists, the effort to understand them and identify conditions for correct use of the component is often as complex as directly investigating the component description. Manually inspecting the VHDL code to retrieve the design intents is a difficult and time consuming activity, especially when the design consists of hundreds or thousands of lines. Moreover, concurrent statements in VHDL introduce a greater degree of complexity to the task, since relations among signals may be spread across different processes. It is evident that new analysis tools which help designers in understanding VHDL source code are necessary.

Goals

In the software engineering area, it has been said [Bigg94] that "*a person understands a program when able to explain the program, its structure, its behavior, its effects on its operational context, and its relationships to its application domain in terms that are qualitatively different from the tokens used to construct the source code of the program".*

In the hardware engineering area, particularly in digital designs, there is no common agreement as to what understanding a component's function consists of. Generally, it can be said that a circuit function is known when its combinational and sequential equations are available in the form of algebraic expressions or finite state machine models. But these representations are too detailed and may be as complex to read as the VHDL code itself.

One of the goals of our project is to identify abstract levels of information which may be used by the designer to understand a component's function and the functional constraints which might limit the component integration into a new design. Specifically, we are developing a software tool, VHDL Assistant Low Efforts Tool (VALET), which assists designers by automatically, or semi-automatically, performing various code analyses on available components. Designer will be able to use the VALET tool to understand a component's function and the functional constraints which might limit the component integration into a new design. Our analysis techniques extract information directly from VHDL code with the goal of extracting functional attributes of the component which is represented by the VHDL code and not information about the VHDL code itself.

Environment

In our project, the reuse scenario consists of a complete VHDL-based design flow, where a repository of VHDL source files of synthesizable components is used. Particularly, each file consists of an RTL description [IEEE99] with its VHDL entity and associated architecture descriptions.

Figure 6.1 shows the integration of the VALET tool into a VHDL-based ASIC design flow. As shown, the tool assists the designer during the translation of a design specification into an RTL design description. In fact by performing various analysis on existing VHDL components, the designer may select an appropriate component or components to be reused.

A repository of VHDL files is directly accessed by our tool. An informal description of the function of each component is available which allows the designer to select the target components to be analyzed.

A graphical user interface (GUI) is provided to allow the designer to explore different parts of a component by selecting design elements (*i.e.* signals, instances and ports) and by invoking different analyses.

Each analysis adds some knowledge about the VHDL code and the described components. The extracted results are reported in the form of graphs and attributes, and are saved in a common information database. Interaction is an important feature of our tool. In fact, it is the designer who guides the tool through the analysis process making appropriate decisions based on the information reported by the tool. Once a designer feels confident about his/her understanding of the component, the associated VHDL code may be extracted from the repository and used in the new design.

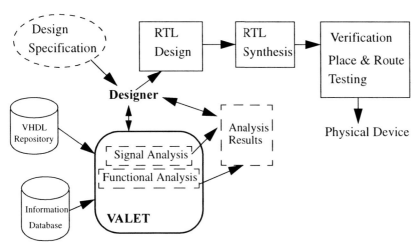

Figure 6.1 VALET tool environment

6.4 FROM VHDL SOURCE CODE TO DESIGN KNOWLEDGE

Since a synthesizable VHDL component always represents a digital hardware circuit, the semantic domain is bounded. Consequently, the problem of extracting design abstractions from VHDL source code is somewhat simplified compared to a similar software reverse engineering problem called the *concept assignment problem* [Bigg93].

It is important to emphasize that there is a limit to the amount of information that can be extracted directly from VHDL code. Knowledge regarding versioning, synthesis scripts, and component history is not accessible from the code. Furthermore, knowledge regarding the specific application domain of the component cannot be inferred from the source code but it must be acknowledged by the designer.

On the other hand, VHDL code contains all the information regarding the actual behavior of a component, including its internal structure and its external interactions. In VALET, using dataflow and control flow data, we are able to extract knowledge at different levels of abstraction and to identify abstract concept models. These models reflect specific behavioral patterns which are typically found in any digital design application. In particular, we have developed analyses which identify behaviors at both the signal and functional levels.

Starting from a VHDL script like the one shown in Table 6.1, a language independent representation called a Finer Program Dependence Graph (FPDG) is built. In this graph, all VHDL code information is maintained and dataflow and control flow dependencies among VHDL statements and data are explicitly represented.

Using the FPDG representation *signal analysis* can be performed. This analysis aims to recognize specific signal use patterns which are typically found in any digital design. Results from the signal analysis allow designers to identify the most important input and output signals as well as the critical internal signals in a design.

```
LIBRARY IEEE;
USE IEEE.STD_LOGIC_1164.ALL;
ENTITY design IS
   PORT ( clock,reset,out_en,write_en,sel: IN   STD_LOGIC;
   data_in    : IN   STD_LOGIC_VECTOR(7 DOWNTO 0);
   data_out   : OUT  STD_LOGIC_VECTOR(7 DOWNTO 0);
   ctrl       : OUT  STD_LOGIC_VECTOR(1 DOWNTO 0) );
END design;

ARCHITECTURE rtl OF design IS
SIGNAL value : STD_LOGIC_VECTOR(7 DOWNTO 0);
BEGIN
   data_out <= value WHEN out_en = '1' ELSE "ZZZZZZZZ";
   main_proc: PROCESS (clock, reset)
      VARIABLE v1 : STD_LOGIC_VECTOR(1 DOWNTO 0);
   BEGIN
         IF reset = '0' THEN value <= "00000000";
      ELSIF clock'EVENT AND clock = '1' THEN
         IF write_en = '1' THEN
            IF (out_en = '0') THEN v1 := "11"; ELSE v1 := data_in(3 DOWNTO 2); END IF;
            IF (sel = '1') THEN ctrl <= v1; ELSE ctrl <= "01"; END IF;
            value <= data_in;
         ELSE
            IF (sel = '0') THEN ctrl <= "00"; END IF;
         END IF;
      END IF;
   END PROCESS;
   ctrl <= "10" WHEN (write_en = '0') AND (sel = '1');
END rtl;
```

Table 6.1 A VHDL design example

Signal analysis results are used in *functional analysis* to identify typical functional behaviors. Behaviors such as encoding, decoding, multiplexer, registers, counters, *etc.* are located by this analysis. By inspecting the list of identified functional behaviors, their usage of signals and their interconnection relations, designers are able to gain some knowledge of the intent of a design. The results of functional analysis are viewed as an abstract representation of part of the original design description.

6.4.1 Finer program dependence graph

In the hardware engineering literature, control flow graphs (CFG) and data flow graphs (DFG) have been widely used to represent the algorithmic characteristics of a VHDL description. In the software engineering literature, the program dependence graph (PDG) [Horw92] is commonly used since it combines dataflow and control dependencies in a single graph. It represents dependencies among statements in a given algorithm and it makes code optimization easier.

In our research, analyses are performed at different levels of detail. Dataflow and control flow dependencies are needed as well as finer grained knowledge of logic operations. In order to keep all this information in a single representation, a new model, called a Finer Program Dependence Graph (FPDG), has been defined. It is based on the PDG graph to which further elements are added. Like an PDG, an FPDG is easy to build when parsing the code. Each implicit or explicit VHDL process in a VHDL architecture is modelled as a FPDG. VHDL processes represent the main computational parts of a VHDL description and they contain all the information which we want to extract.

A FPDG graph is composed of the following element types.

- *region node*: the root of a sub-tree whose elements have common control dependencies. A region node represents the entry point for a set of sequential statements which depend on the same condition. These nodes divide the process into sub-areas which can be individually analyzed and then combined. An attribute is associated with a region node to indicate whether the elements of the associated sub-tree belong to a clocked portion of the design.

- *predicate node*: represents a conditional statement in the process. The original expression and an equivalent sum of product expression are associated with each predicate node. These two representations support different types of analysis.

- *statement node*: represents an unconditional statement, *i.e.* a signal assignment, a variable assignment or a procedure/function call. A statement node contains a sub-graph which explicitly describes the data dependence relations among signals, variables and constant values. In this way, all the fine detail available in the source code is maintained.

- *control dependency edges*: connect nodes of the types above such that the control flow described by the process is captured in the graph. Following these edges through the graph it is possible to identify the conditions for which specific assignments are activated.

- *data dependence edges*: connect nodes based on dataflow information in the process. Connections between nodes in a sub-graph and/or upper level nodes may exist depending on the data dependence relations. In this way, coarse as well as fine dataflow dependencies are explicitly available.

The FPDG representation is language independent so VHDL, Verilog, or other HDL descriptions employing the synthesizable subset of the languages, may be parsed and translated to a set of FPDG representations. In our project only VHDL descriptions are considered.

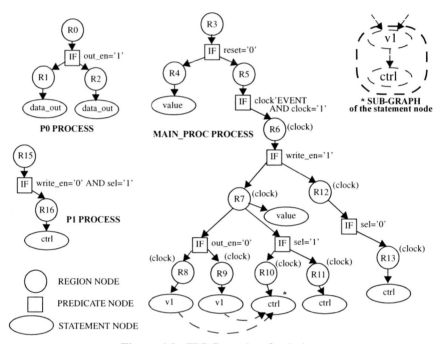

Figure 6.2 FPDG graphs of a design

In Figure 6.2, the FPDG representations obtained by parsing the VHDL code in Table 6.1 are depicted. Processes *P0* and *P1* correspond to the first and second concurrent conditional assignments used in the architecture description. Control dependence edges are identified by solid arrows while data dependence edges by dashed arrows. When more than one dependence edge exits in a region node, a left to right control flow dependence exists. That is, the sequential nature of a VHDL process execution is implicitly maintained by the order of the control dependence edges exiting a region node.

In Figure 6.2, one of the statement nodes (marked *) has been expanded to show its sub-graph. In this sub-graph, the dependence between the *ctrl* signal and the *v1* variable is explicitly indicated by a data dependence edge. In the graph, two data dependence edges are used to indicate the dependencies between this statement node and two other statement nodes where the variable *v1* is assigned.

During the parsing of the code, it was recognized that the signal *clock* represents a digital clock using the pattern rules described in IEEE P1076.6 standard

[IEEE99]. Hence the clock attribute has been assigned to region nodes R6,R7,R8,R9,R10,R11,R12,R13.

By traversing the FPDG, it is possible to retrieve all the information required to perform various analyses available in the VALET tool.

6.4.2 Signal analysis

Signal analysis is the first of our analyses. The purpose is to help designers in identifying uses of signals in the VHDL description at the semantic level. By traversing the FPDG of each process in an architecture, our analysis algorithm is able to automatically categorize each signal based on its use in the VHDL code. In particular, we categorize each signal at two different levels: at the semantic level and at the code level.

For the semantic level, we have identified signal use patterns which are commonly found in a digital design. These patterns consist of relations among signals in which one acts as the controller, another as the controlled signal, while other signals identify the conditions on which the controller is activated. The currently defined signal use patterns are:

- *selector*: For all possible values of a signal (controller), another signal (controlled) is assigned to values. A condition may exist which activates the controller for all its values.

- *partially assigned*: For a subset of values of a signal (controller), another signal (controlled) is assigned to values. A condition may exist which activates the controller for the subset of its values.

- *set/reset*: For a specific value, or a range of values, of a signal (controller), another signal (controlled) is assigned to a specific value. For all other values of the controller signal, different values than the one noted are assigned to the controlled signal or there is no assignment. A condition may exist which activates the controller for all its values.

- *enable*: For a specific value, or a range of values, of a signal (controller), another signal (controlled) is assigned to a value. For all other values of the controller signal there is no assignment to the controlled signal. A condition may exists which activates the controller for all its values.

- *tristate enable*. For a specific value, or a range of values, of a signal (controller), another signal (controlled) is assigned to values. For all other values of the controller signal, high impedance is assigned to the controlled signal or there is no assignment to it. A condition may exist which activates the controller for all its values.

- *tristate disable*. For a specific value, or a range of values, of a signal (controller), another signal (controlled) is assigned to high impedance. For all other values of the controller signal, values are assigned to the controlled signal or there is no assignment to it. A condition may exist which activates the controller for all its values

90   VIRTUAL COMPONENTS DESIGN AND REUSE

The above list does not represent all possible signal behaviors but it includes a reasonable set of commonly used signal patterns.

In some cases, a signal might not be categorized by any of the above patterns or the designer might want to further investigate its use by examining the VHDL code. In order to assist the designer in such activity, our algorithm identifies the region nodes, with their dependence sub-tree, in which a signal satisfies one of the following conditions:

- *sampled*: The signal is used only in statement nodes that depend on the region node, and it is assigned to other signals with no logical, arithmetic or shift operations except, possibly, for the logic NOT operator.
- *encoded*: The signal is used in all statement nodes which depend on the region node, and only constant values are assigned to it.
- *status encoded*: The signal is used in some but not all statement nodes which depend on the region node, and only constant values are assigned to it.
- *choice*: The signal is used in all statement nodes which depend on the region node, and only signals with no logical, arithmetic or shift operation except, possibly, for the logic NOT operator, are assigned to it.
- *status choice*: The signal is used in some but not all statement nodes which depend on the region node, and only signals with no logical, arithmetic or shift operation except, possibly, for the logic NOT operator, are assigned to it.

A designer can use the above information to reduce the lines of VHDL code to investigate. For example, by focusing on the VHDL code which is represented by the sub-tree of an identified region node where a signal is always assigned to a constant value, i.e. an *encoded* condition. In Table 6.2, a sampling of the results obtained by our signal analysis algorithm for the VHDL description of Table 6.1 are reported. For each signal the table shows the role of the signal in a semantic signal use pattern and/or its code level usage. Signal use pattern conditions are reported in brackets. Note that the table is not complete, particularly for the signal *ctrl*. No details on condition and dependencies are reported and for the *reset* signal only a limited list of usage patterns are shown. Furthermore clock dependencies are not explicitly shown.

From the table it can be noticed that a signal may belong to more than one category or to the same category but with different conditions. For example, consider the *write_en* signal categorization. Under the condition *reset='1' and sel='1' and out_en='1'*, it acts as a *set/reset* by controlling signal *ctrl*. In fact, under that condition, the algorithm identified that when *write_en* has value '0', the assignment *ctrl<="10"* is executed in process *P1*, while when *write_en* has value '1' the assignment *ctrl<=data_in(3 to 2)* is executed in process *main_proc*. The assignment *ctrl<=data_in(3 to 2)* has been obtained by considering dataflow dependencies between *ctrl* signal and *v1* variable. Note that, even though not explicitly shown in the table, VALET informs the designer that such signal behavior is subject also to the rising edge of the clock.

Signal	Semantic level	Code level
out_en	controller for data_out in *tristate enable*	
write_en	controller for ctrl in *set/reset* (reset='1' and sel='1' and out_en = '1') controller for ctrl in *selector* (reset='1' and sel='1' and out_en = '0') controller for ctrl in *selector* (reset='1' and sel='0') controller for value in *enable* (reset='1')	
sel	controller for ctrl in *selector* (reset='1' and write_en='0') controller for ctrl in *set/reset* (reset='1' and write_en='1' and out_en='1') controller for ctrl in *selector* (reset='1' and write_en='1' and out_en='0')	
reset	controller for value in *set/reset* (write_en='1') controller for ctrl in *enable* (write_en='0' AND sel='0') ...	
ctrl	controlled by sel in *selector, set/reset* controlled by write_en in *set/reset, enable* and *selector*	*encoded status* in R12
data_out	controlled by out_en in *tristate enable*	*choice* in R1
data_in		*sampled* in R3
data_in(2 to 3)		*sampled* in R3
value	controlled by write_en in *enable* controlled by reset in *enable*	*choice status* in R5

Table 6.2 Signal analysis results

Under condition *reset='1' and sel='1' and out_en='0'*, the algorithm identifies a *selector* pattern where *write_en* control the assignments *ctrl<="10"* in process *P1* and *ctrl<="11"* in process *main_proc* considering dataflow dependencies. The *write_en* signal is also the controller in another *selector* pattern under the condition: *reset='1' and sel='1'*. In addition to controlling the *ctrl* signal, the *write_en* signal controls the *value* signal in an *enable* pattern under the condition: *reset='1'*. In this case, the algorithm identifies that when *write_en* has the value '1', a value is assigned to the *value* signal, while for *write_en*='0' no assignment exists for the *value* signal.

The results of code level signal analysis are not very enlightening for the given example since most of the information is already given by the semantic level analysis. However, as an example, consider the result: *ctrl* signal in region R12 has an *encoded status* condition. This information underlines that, considering only the *main_proc* process, there is at least one operating condition in which the *ctrl* signal maintains its value between two different executions of the *main_proc* process.

As shown in the example, signal analysis takes into account dataflow dependencies among signals and variables and it identifies usage patterns among multiple processes.

6.4.3 Functional analysis

Signal analysis finds relations between controlling and controlled signals which can be used by the designer to recover part of the original design intent. *Functional analysis* aims to determine more complex interactions amongst signals such that higher-level functional abstractions can be identified.

We have to date studied a set of basic functional behaviors which can be identified by functional analysis. Our set is not complete, that is it does not contain all possible hardware behaviors, so only partial functional recognition is achieved. It means that we may not recognize the entire VHDL description of a component as an instance of one single function, but rather as an aggregate of various functional behaviors.

Currently the following functional behaviors have been defined and are recognized by our VALET tool:

- *encoder*: There exists a set I of signals and a set O of signals. For each possible value over the set I, a constant value is assigned to each signal in O. The number of bits used by all signals in I is greater or equal to the number of bits of each signal in O.

- *decoder*: There exists a set I of signals and a set O of signals. For each possible value over the set I, a constant value is assigned to each signal in O. The number of bits used by all signals in I is less then the number of bits of each signal in O.

- *multiplexer*: There exists a set S of signals and a signal o. For at least k values over the set S, the value of o is equal to a signal $i \in I$ with $|I|=k$. The value k is defined by a designer using a parameter md which we call *multiplexer degree*, $k = md \times values(S)$, where *value(S)* is the number of possible values over S. For md equal to 1, the functional analysis tries to find a complete multiplexer behavior where for each value of S a different signal is assigned to o. A designer may freely choose md, as much as k is at least greater or equal to 2, to control multiplexer behaviors identified by VALET.

- *demultiplexer*: There exists a set S of signals and an signal i. For at least k values over the set S, the signal i is assigned to a signal $o \in O$ with $|O|=k$. The value k is defined by a designer using a parameter dd which we call *demultiplexer degree*, $k = dd \times values(S)$, where *value(S)* is the number of possible values over S. For dd equal to 1, the functional analysis tries to find a complete demultiplexer behavior where for each value of S, the signal i is assigned to a different signal o. A designer may freely choose dd, as much as k is at least greater or equal to 2, to control demultiplexer behaviors identified by VALET.

- *parallel register*: In its simplest form, it consists of two signals: i and o composed by more than one bit. Signal i is assigned to signal o depending on the rising or falling edge of a clock. Using the results from signal analysis, functional analysis is able to further categorize such functional behavior by eventually associating working conditions, enable signal and reset/set signal.

Functional analysis is performed using results from signal analysis together with information about dependencies among identified signal patterns.

For the VHDL design in Table 6.1, the functional analysis results are shown in Table 6.3. From signal analysis results, *write_en* could have been directly considered as the input to a decoder with output *ctrl* and its activation condition depending on *sel*. Vice versa, *sel* could have been directly considered as input to a decoder with output *ctrl* and its activation condition depending on *write_en*. As shown, our algorithm is able to automatically combine signals *write_en* and *sel* to form the input of an encoder functional behavior with the condition reset='1'. In fact, when *sel*='0' and *write_en*='0' then *ctrl*<="00"; when *sel*='0' and *write_en*='1' then *ctrl*<="01"; when *sel*='1' and *write_en*='0' then *ctrl*<="10"; and when *sel*='1' and *write_en*='1' then *ctrl*<="11".

Functional behavior	Details
encoder	With set I composed of signals *write_en* and *sel*; and set O of signal *ctrl*. Condition *reset*='1' exists.
parallel register (I)	With *data_in* as i signal and *value* as o signal. Signal *reset* as set/reset, signal *write_en* as enable.
parallel register (II)	With *data_in(3 downto 2)* as i signal and *ctrl* as o signal. Signal *sel* as set/reset, and condition *reset*='1' and *out_en*='0'.

Table 6.3 Functional analysis results

Regarding parallel register identification, from the signal analysis *data_in* and *value* are immediately identified as the *i* and *o* signals of a parallel register, respectively. The dependencies for the *value* signal in the set/reset pattern with *reset* as controller and in the enable pattern with *write_en* as controller allows our algorithm to specialize the parallel register with *reset* and *write_en* as respectively reset/set and enable for the register. Similarly, *data_in(3 dowto 2)* and *ctrl* signals are recognized as belonging to a parallel register with signal *sel* as an enable for the register.

Note that the goal of functional analysis is to locate functional behaviors which might not be synthesized as such in the actual digital implementation by a synthesis tool but which are useful for understanding the overall function of the design. For this reason, the clock rising or falling condition is always implicitly maintained but not explicitly used as a condition.

6.5 CONCLUSIONS

In this article we have described our approach to assisting designers in understanding VHDL legacy code. Understanding VHDL code is one of the activities which is required in the reuse of existing designs. We have described an incremental analysis methodology in which different analyses are performed in order to gather information at differing levels of abstraction directly from VHDL code.

Our methodology has been implemented in a software tool called VALET. The interaction with a typical IC design flow has been depicted. Examples of analysis results which have been obtained using VALET have been reported. The algorithmic details of VALET have not been reported due to space constraints. We invite readers to contact the authors if interested.

The VALET project is still under development and improvements and new analyses are continuously being added. We are currently working on a new algorithm which enlarges the capability of functional analysis by taking into account more dataflow and inter-process relations among identified signal usage patterns. For example, referring to results on tables Table 6.2 and Table 6.3, the new algorithm should combine the *tristate enable* pattern for controlling *out_en* signal and controlled *data_out* signal with the parallel register functional behavior, which uses *data_in* and *value* signals, to recognize that the signal *out_en* is an enable controller for the register *value* signal. Algorithms to handle counter behavior and finite state machine transitions are also under investigation.

Even though the currently developed analyses are only able to achieve limited results, they indicate a new way to assist designers in the understanding and reuse of legacy designs.

7 LAMBDA-BLOCK ANALYSIS OF VHDL FOR DESIGN REUSE

William Fornaciari, Salvatore Minonne, Fabio Salice and Massimo Vincenzi

Politecnico di Milano
Dipartimento di Elettronica e Informazione
Milano, Italy

ABSTRACT

The chapter presents a methodology to be used for both design and analysis of digital systems described by using VHDL. The developed CAD environment allows the designer to inspect the code of existing systems in order to extract candidate functionality suitable for reuse as well as to evaluate the quality of VHDL in a reuse perspective. The proposed methodology has been validated by considering small industrial benchmarks and by redesigning an industrial core cell, a PC-Card interface, used in commercial devices.

7.1 INTRODUCTION

Improvements in silicon capabilities are hard to be followed by EDA (Electronic Design Automation) methodologies and tools evolution. The mismatch between the designer's productivity and the growth rate of the device sizes (the so-called productivity gap) is becoming more and more evident. Semantech reports that in the last five years the silicon capability increased by 50% per year, while the designer's productivity increases only by 21% [Rass99]. To sustain such a steady progress, the entire management of the design flows should be reconsidered, in particular because the time-to-market pressure is expected to shrink the productivity cycle of realizations in between 6 and 9 months in the next three years.

To cope with such problems, the viable solution seems to be the emphasis on standards (protocols, HDLs, interchange formats for tool interoperability, ...) together with a new design style taking into account the possibility of incorporating existing components in new designs. Reusability and IP-based design are hot topics for both the industrial and academic worlds, and new important initiatives started, such as the VSIA [VSIA00b] and the RASSP [Rass99] program. Third-part suppliers of IP cells are now emerging on the market arena.

Industries started considering the problem of reuse a long time ago, and currently they are investigating the cross relations existing between standards, EDA design flows and costs. The goal is to discover the driving forces for the make-or-buy decision. For instance, Siemens [Prei95a] found that the design of a new fully parametric component suitable for reuse could be up to four times longer than the ad-hoc design. A trade-off analysis performed by Synopsys [Keat98b] asserts that the design effort for realizing a new component targeted for external reuse is three times longer, while for internal reuse the design time is only doubled in comparison with a committed design. Other successful experiments [Gast98] have been carried out by considering a 800kgates system, designed by reusing existing parts: 20% consisted of a netlist and 30% of VHDL-RTL.

A common feeling among the designers is that the effort for reusing components must be below 30% of a starting from scratch design in order to be cost-effective [Girc93]. For this reason, it is of paramount importance to consider other factors in addition to the purely functional requirements affecting the understandability of a design, such as the quality of documentation, the VHDL coding style and the adaptability of the implemented core functionality.

The long-term goal of the research we are performing, as explained also in the Section 7.2, presenting the overall environment, is twofold:

- Set-up of a business model to predict the cost-effectiveness of internal reuse
- Identification of set of metrics, capturing the degree of reusability of already designed VHDL-RTL components

The focus of this contribution is on the reuse activity: definition of an analysis methodology providing metrics to estimate the effort necessary to reuse existing VHDL designs, and identification of candidate portions of VHDL code containing basic functionality.

Section 7.3 and Section 7.4 outline the main concepts considered to identify reusability measures: testbench orthogonality and λ-*block*. The former considers the testbenches as a significant part of a design and analyzes their capability to cover specific functionality of the overall system. The latter identifies blocks of VHDL constructs to highlight properties similar to those captured by basic blocks of software-oriented formalisms.

In Section 7.5, the main results obtained by applying the proposed analysis methodology onto such a design are reported, since it is a meaningful example of industrial device conceived for reuse. Concluding remarks are drawn in Section 7.6, which also outline the focus of our current investigation effort.

LAMBDA-BLOCK ANALYSIS OF VHDL FOR DESIGN REUSE

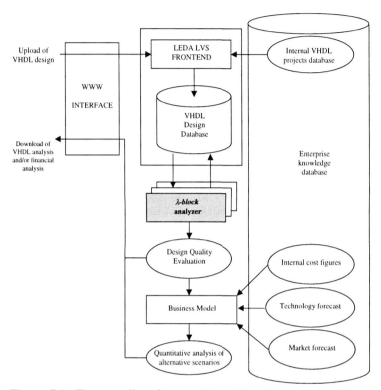

Figure 7.1 The overall environment to analyze design reusability

7.2 OVERVIEW OF THE GLOBAL ENVIRONMENT

The investigation presented, is part of a bigger project aiming at providing an analysis framework to evaluated the cost-effectiveness of reuse (see Figure 7.1). The methodology is attacking the problem from two sides: VHDL-related issues and modelling of the potential financial advantages. We are considering the reuse for internal purposes, i.e., the design of VHDL components to be (re)used as well as the resuming of existing VHDL projects, constituting in many cases a significant intellectual (underused) capital of the company.

Time-to-market pressure and designer's mobility rise up the importance of following design guidelines to simplify teamwork, component reuse and overall quality. The proposal presented in the next section (grey box in Figure 7.1) is concerning only a part of the proposed methodology. The modelling of a VHDL-RTL design and an analysis methodology is outlined, to capture potential basic functionality candidates for the reuse as well as to quantify the complexity of understanding (and thus modifying) of the original VHDL code. The analysis also allows the evaluation of some important characteristics of the VHDL code quality, such as dangling signals, signals defined but never used, etc. Currently, work is in progress aiming at extending the set of implemented analysis strategies which, in turn, will allow the designer to produce financial forecasts and/or to verify significant design properties.

An experimental analysis EDA tool has been implemented on top of the LEDA LVS environment [Leda00], providing a front-end and a suitable VHDL database IEEE-1076 compliant. LVS performs a complete parsing of VHDL and exposes an internal database structure suitable for inspection and manipulation. The result of the analysis process can be employed as a set of the metrics used to estimate the complexity of design reuse within a business model, and in parallel, a pure text output is returned or web-based output is made available. Also this capability is under development in order to simplify the managing of projects though a company Intranet. In such a way, the tool becomes also accessible to external users, having only to upload their files, tune the analysis parameters and get back the results on their web browser, while preserving the security of the company design database.

7.3 TESTBENCH ORTHOGONALITY

The typical activity of a VHDL designer is related to the use of a number of templates forced ba the application of synthesis tools, readability of code and the presence of specific design goals (e.g., low power, speed, etc.). Especially for lack of documentation, the resulting RTL is not easy to follow and problems like buffering of signals in case of high fan-out are hard to be recognized and managed during reuse. Testbenches are also very important for reuse, and therefore, they should be conceived before the actual component design; their main characteristics are:

- They contain code and in particular data to be used for RTL testing.
- They enable a comprehensive RTL simulation, simplifying the localization of errors. It is also possible to introduce specific information for timing and functionality.
- They provide the designer with a better understanding of the component behavior, because they usually describe the same functionality at a higher abstraction level.

We introduce the concept of othogonality associated with a VHDL entity. A functionality is identified as an input/output relation verified by one or more testbenches. In turn, the orthogonality is an attempt to predict the partitionability of a component, since it allows:

- The identification of a given functionality; this aspect is important at design time to find out the functional blocks as well as to identify a proper testing strategy.

- The improvement of the reuse of functions, since their extraction from the first design is simplified.

7.4 LAMBDA-BLOCKS ANALYSIS

As far as the analysis of the VHDL functionality is concerned, and similarly to sequential languages it is possible to identify blocks of instructions relevant for the understanding of the specification behavior. Unfortunately, this activity is more complicated since in VHDL the information related to the computation is, at any point in time, influenced by the values of the signals, i.e. the concept of *current instruction* is not applicable. For this reasons, we introduced a computational abstract model resembling the λ-calculus of the functional languages. The concept of basic block has been reconsidered by focusing the attention on a subset of the concurrent statements to identify portions of code (λ-blocks) relevant for static analysis of the functionality. The assumption is that λ-blocks analysis allows to expose the template used by the original designer, which can simplify the understanding of components during reuse and, by considering the connectivity among such blocks, to concisely model the potential data flow within the device.

The concurrent constructs considered by λ-blocks are: "process", "concurrent procedure call", "when-else", "when-select" and "component" instantiation. Three constructs have been excluded in the first implementation (assert, block, generate) since they do not actually impact on the functionality, being only a methodological design support. Since the `generate` constructs are not considered, lambda analysis has to be carried out when all the cross-references among design units are removed.

As cited above, a λ-blocks determine any functionality, with some inputs and outputs, capturing the interconnections among concurrent constructs. For example, the inputs of a λ-blocks associated with a "process" contain the following signals:

- Signals or ports belonging to the sensitivity list of the process and of possible wait-on.
- Signal or ports in the right-hand part of an assignment, or that are actual parameters associated with formal IN/OUT parameters of functions (or procedures) in the right part of an assignment.
- Signal and ports within a conditional expression.

The λ-blocks outputs of a "process" are signals or ports in the right part of an assignment.

The "when-else" construct, including signal assignment as a special case, has the following inputs:

- Signals or ports in the right part of the constructs, or that are actual parameters ("IN" or "INOUT") of a function or procedure located in the right side.

- Signals or ports located within the conditional expression or that are formal parameters of a function or procedure contained by the expression.

For example, the expression:
```
output <= a when b='1' else c;
```

has the following Input (I) and Output (O) sets:
I=<a,b,c>, O=<output>

The "with-select" construct is similar to "when-else":
- The inputs are the signals or ports on the right, as well as those composing the condition.
- The assigned signals or port constitute the output set.

For example, the input and output sets of the following code are I=<a,b,c, sel>, O=<output>.
```
with sel select;
   output <= a when ''00''
      b when ''01'',
      c when ''11'',
      d when others;
```

Concerning component instantiation, the input set includes the actual parameters associated with the modality "IN", "INOUT", "BUFFER", while the output set consist of those "OUT", "INOUT" and "BUFFER".

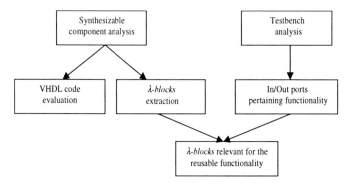

Figure 7.2 The lambda-block analysis flow

The use of λ-blocks analysis is depicted in Figure 7.2. Once λ-blocks have been extracted and testbench analysis performed, it is possible to identify the λ-blocks activated by a functionality identified through a testbench. This activity is carried out in three steps:

Let us consider the example reported in Figure 7.3, where we suppose the availability of a testbench using only the input port "I1". By forward propagating the data, a cone of λ-blocks from inputs to outputs are identified (grey bubbles). Note that λ-b5 and λ-b6 are not activated, since they are reachable only starting from "I2".

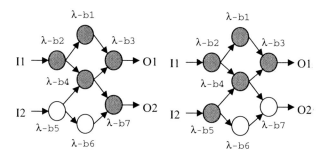

Figure 7.3 Forward activation (left) and backward propagation (right)

Furthermore, by assuming that output "O1" is relevant for the functionality we are going to identify, a backward propagation of such an information can be carried out, as shown in Figure 7.3 (right), and a new cone of λ-blocks is identified. Eventually, the intersection of the cones allows highlighting only the relevant λ-blocks, as reported in Figure 7.4.

Suitable metrics are used to select the best set of λ-blocks ensuring the coverage of the entire system structure. These metrics allow to evaluate the complexity of VHDL and the effort for its reuse. They are based on the analysis of the λ-blocks count, the frequency of each λ-blocks and the minimum/maximum *depth* of λ-blocks.

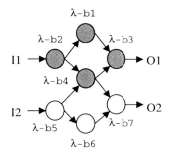

Figure 7.4 Functionality extraction via intersection of λ-blocks

λ-blocks analysis also allow to inspect the quality of VHDL code, in terms of the following aspects:

102   VIRTUAL COMPONENTS DESIGN AND REUSE

- Signal dangling: λ-blocks outputs which are neither inputs of other λ-blocks nor outputs of the component.

- Signal without driver: λ-blocks inputs which are neither inputs of the components nor output of other λ-blocks.

- Signals defined and never used: signals that are neither inputs nor outputs of λ-blocks.

7.5 THE REUSE EXPERIMENT

The methodology and the environment have been evaluated by analyzing a real design, a PCMCIA soft-core interface component, which is a natural candidate for reuse since it implements a standard. After the redesign of the overall interface to remove any dependency with the original use context, the analysis revealed a number of λ-blocks. This was one order of magnitude smaller than the number of VHDL lines. (cf. Table 7.1)

VHDL Lines	1180
λ-blocks	120
When Else	93
Process	22
Component Instantiated	5

Table 7.1 Analysis of the VHDL concurrent statements

This analysis shows that the coding style is focused on the "when-else" construct instead of "process". By applying the testbench analyzer tool, four different groups of partially overlapped λ-blocks have been identified. (cf. Table 7.2)

The mismatch between the total number of λ-blocks identified after the analysis means that the analyzed functions are not completely orthogonal, i.e. some logic is shared. The analysis is rather precise, which can be checked by comparing the columns of Table 7.2 (error below 5%).

The application of the tool is a significant help for the *reuse-oriented* designer, since it allows the immediate location of the main functionality, and the possibility to work with description sizes that are around half of the total description. It was possible to identify all the dangling signals, constituting 13% of the total.

Table 7.3 shows the distribution of the effort between the different activities performed during the experiment. As it points out, a relevant part of the effort has been devoted to understand the PC-Card standard, while the adopted methodology significantly improved VHDL-related activity.

Functionality	λ-blocks extracted	λ-blocks actually pertaining the functionality
Configuration	61	58
Memory R/W	71	70
I/O	71	70
EEPROM Configuration	34	33
TOTAL	237	231

Table 7.2 Lambda-blocks extracted and grouped by functionality

Activity	Effort
Understanding of PC-Card standard	50%
Extraction of PCMCIA core and pruning of the original customizations	6%
Compilation and functional debug	12%
Post-synthesis simulation	30%
Testing and final simulation	2%

Table 7.3 Manpower distribution

7.6 CONCLUDING REMARKS

This contribution presents an overview of a methodology to analyze VHDL designs in order to evaluate their suitability to be reused and to extract basic functionality. The main concepts on which the investigation is built (and metrics have been derived) are testbench orthogonality and λ-blocks analysis.

The actual implementation still suffers of some limitations. In particular, the amount of λ-blocks identified can be bigger with respect to those stimulated during actual simulation. This can happen since static information are considered (e.g., some branches will never be executed) and some signals (e.g. clock, reset) are inputs of λ-blocks even if they are not relevant in terms of functionality extraction. However, the case study considered up to now, shows that the obtained accuracy is widely acceptable. The proposed analysis revealed his suitability during a reuse-experiment of a commercial device. It improves the quality of the specification, simplifies the identification of design bugs and it insulates the basic functionality composing the specification.

Work is in progress to define a framework estimating the cost effectiveness of internal reuse by considering the metrics. The web-accessible model will also take into account typical industrial design flows, identifying the overall additional costs and potential savings.

8 IP RETRIEVAL BY SOLVING CONSTRAINT SATISFACTION PROBLEMS

Manfred Koegst*, Jörg Schneider*,
Ralph Bergmann** and Ivo Vollrath**

*Fraunhofer-Institute for Integrated Circuits,
Department EAS,
Dresden, Germany

**University of Kaiserslautern,
Department of Computer Science,
Kaiserslautern, Germany

Abstract

The increasing productivity gap implies the necessity to integrate reuse approaches in the circuit and system design. To install reuse approaches in the design flow, adapted database systems are needed for managing the reuse process, and especially for archiving and retrieving of intellectual properties (IPs). Our contribution is focused on the IP retrieval task. It builds upon two recent technologies for building knowledge-based systems: case-based reasoning and constraint satisfaction. For implementing this retrieval approach we develop a concept of an adapted database management system for archiving IP specifications utilizing a structured representation of relations between IP attributes.

8.1 INTRODUCTION

Beside area, timing, and power, time-to marked has become more and more an important optimization parameter in the design process of ICs. The reuse of existing and validated components (intellectual properties, IP) can contribute to enhance design efficiency as desired. Consequently, related reuse-based design methodologies have to be integrated in industrial design flows [Reut99]. First experiences in applying reuse approaches in the industrial circuit design are presented e.g. in ([Haas99b] [Mous99] [Reut99]). Some methodologies were discussed in [Koeg98] utilizing the model of a reuse cycle. One of the main tasks is the retrieval of parametrized IPs according to requirement specifications, where retrieval is performed utilizing an adapted database management ([Faul99] [Haas99a]).

Retrieval of IPs is a crucial subtask ([Berg99a] [Oehl98]), which can be supported by case-based reasoning (CBR) ([Berg99a] [Berg99b]) and constraint satisfaction ([Meye96] [Tsan93]) technologies, known from Artificial Intelligence. Based on these technologies, retrieval is modelled and a new approach for IP retrieval is proposed. Finally, an adapted database management is proposed for implementing this approach considering the hierarchy of attribute domains and of the related requirements

Our contribution is structured as follows: In Section 8.2, we describe the archiving and retrieval task in the case-based reasoning terminology of Artificial Intelligence. The aim of Section 8.3 is to model IP retrieval by case-based reasoning. We define a generalized query by a set of constraints ([Meye96] [Tsan93]). After classifying constraints we model classes of binary constraints by matrices and perform problem solving by matrix multiplication. In Section 8.4, we describe the conception of our database structure for IP archiving and representation of IP attributes. Additionally, in Section 8.6 the introduced framework is illustrated by means of a small example.

8.2 ARCHIVING AND RETRIEVAL IN THE REUSE PROCESS

Case-based reasoning (CBR) [Berg99b] is a recent approach for building knowledge-based systems. The idea of CBR is to avoid solving problems from scratch by applying and combining general knowledge but to reuse specific previous problem solving experience to improve the problem solving process for future problems. Problem solving experience is captured in a set of so-called *cases* stored in a *case base*. A case records a previous problem together with its solution. A case can be formalized as a point (x,y) in the problem-solution space $\underline{X} \times \underline{Y}$ where $\underline{X}$ is the set of problems and $\underline{Y}$ the set of solutions.

Input for the classical case-based problem solving is a new problem x' named *query*, and problem solving is the task to select a case (x,y) from the case base with the property that query x' and problem x are equal or similar concerning a specified similarity function.

When applying the CBR idea to IP reuse, the solution space $\underline{Y}$ corresponds to the set of all available IP specifications and the problem set $\underline{X}$ corresponds to the set of all values of IP attributes. Hence, a case (x,y) characterizes a relation between a single attribute vector x and a single IP specification y. However, for describing parametrized IPs and IP classes, this formalization is not sufficient. Therefore, we build upon a recently proposed extended case representation called *generalized case* [Berg99b]. A generalized case is defined as a subspace (X,Y) of $\underline{X} \times \underline{Y}$ instead of a point.

Let $V = \{v_1,...,v_\rho,...,v_r\}$ be the set of attributes of the IPs and X_ρ the domain of attribute v_ρ. $\underline{X}$ can be specified by the Cartesian product $\underline{X} = X_1 \times ... \times X_\rho \times ... \times X_r$. For a subset A of V, we describe the corresponding subspace of $\underline{X}$ by $\underline{X}_A = \times (X_\rho \mid v_\rho \in A)$ assuming that the components in this Cartesian product are ordered by the indices of their attributes. In [Koeg98] a *constraint* in space $\underline{X}$ is defined by a pair $c = (A,B)$ with $\emptyset \subseteq A \subseteq V$ and $\emptyset \subseteq B \subseteq \underline{X}_A$, where A is a set of attributes whose values are constrained, and B is the set of allowed value vectors. For $B = \emptyset$ constraint c is empty, described by $c = 0$. For $B = \underline{X}_A$ constraint c is no restriction of the space $\underline{X}$. In the terminology of [Berg99a] constraints are used to represent a generalized problem, i.e., to a part of the problem space $\underline{X}$. In this approach a set of constraints is used to present a query. A query is empty if it contains excluding each other constraints.

For *archiving* IPs in a library, it is to assign to an IP specification y a related constraint x (e.g. key words, technical parameters) necessary for retrieving this IP (Figure 8.1). This assignment corresponds to a mapping

$$f: \underline{Y} \rightarrow \underline{X}$$

which is not unique because different IPs may share the same constraints.

Formally, *retrieval* is the inversion of archiving, but in practice it is performed in two steps characterized by a pair (g_1, g_2) of mappings. Because f is not a unique mapping its inversion has to be a relation between the corresponding power sets

$$g_2: \wp(\underline{X}) \rightarrow \wp(\underline{Y})$$

where each pair (X,Y) with $X \in \wp(\underline{X})$ and $Y \in \wp(\underline{Y})$ is a generalized case in the case-base. Such a generalized case (X,Y) corresponds to a class of IPs. For *retrieving* an IP from a design library, a user formulates its requirements usually not by a query X but by a set C of constraints between IP attributes. The constraints of set C can restrict each other or can even be inconsistently. Therefore, in a first retrieval step the user-specified constraints C must be transformed into a subspace X. This corresponds to a mapping

$$g_1: \wp(\underline{C}) \rightarrow \wp(\underline{X}),$$

where $\underline{C}$ denotes the set of all possible constraints concerning set V of attributes. In this representation mapping g_1 describes the *problem reduction* achieved by assigning IPs to the constraints (Figure 8.1).

In the next chapter we model the main steps of retrieval, namely problem reduction and problem solving by case-based reasoning and solve them by matrix operations.

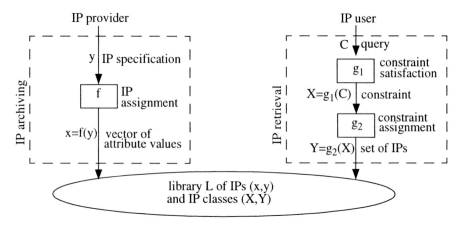

Figure 8.1 Scheme for IP archiving and retrieval

8.3 MODELLING AND RETRIEVAL BY CASE-BASED REASONING

8.3.1 Constraint relations

The basic term for specifying user requirements and queries for retrieval are constraints $c = (A,B)$, as defined in the previous section. Following the database terminology (e.g. [Heue97]) we introduce the relations *projection, join,* and *satisfaction* for constraints. Concerning a constraint $c = (A,B)$ we characterize by $v_\rho(x)$ the value of attribute v_ρ in vector x, where v_ρ and x are elements of A and B, respectively.

E.g., for constraint $c = (A,B)$ with the set $A = \{v_1, v_2, v_4\}$ of attributes and the set $B = \{010, 110, 011\}$ of vectors the value of attribute v_2 in vector 110 is given by $v_2(110) = 1$.

Definition 1: The *projection* $B' = (c \bot A')$ of constraint $c = (A,B)$ to a subset A' of attributes with $\emptyset \subset A' \subseteq A$ is the reduction of all vectors of set B concerning the attributes of A', defined by

$$x' \in B' \text{ iff } x' \in \underline{X}_{A'} \text{ and } \exists\, x\, (x \in B \land \forall v_\rho\, (v_\rho \in A' \rightarrow v_\rho(x) = v_\rho(x'))).$$

E.g., the projection $B' = (c \bot A')$ of constraint $c = (\{v_1, v_2, v_4\}, \{010, 110, 011\})$ to set $A' = \{v_2, v_4\}$ is $B' = \{10, 11\}$.

Definition 2: The *join* of two constraints $c' = (A',B')$ and $c'' = (A'',B'')$ is a constraint $c = (A,B)$, described by $c = c' \oplus c''$, with the attribute set $A = A' \cup A''$ and where each vector x of set B is a combination of a vector x' of B' and a compatible vector x" of B", defined by

$$x \in B \text{ iff } x \in \underline{X}_A \text{ and } (A,\{x\}) \bot A') \subseteq B' \text{ and } (A,\{x\}) \bot A") \subseteq B".$$

E.g., the *join* of the constraints $c' = (\{v_1,v_2\},\{01,11\})$ and $c" = (\{v_2,v_4\},\{00,10,11\})$ results in constraint $c = (\{v_1,v_2,v_4\},\{010,110,011,111\})$.

Join is an associative relation, i.e. it satisfies the law

$$(c \oplus c') \oplus c" = c \oplus (c' \oplus c") = c \oplus c' \oplus c".$$

For *joining* a set C of constraints c we use the abbreviation $\oplus (c \mid c \in C)$.

Definition 3: A constraint $c = (A,B)$ satisfies constraint $c' = (A',B')$ iff $A' \subseteq A$ and the disjunction of B' with the projection of c to A' is not empty, i.e., $((A,B) \bot A') \cap B' \neq \emptyset$ (generalization of [Tsan93]). E.g., constraint $c = (A,B)$ with $A=\{v_1,v_2,v_3,v_4\}$ and $B=\{1100\}$ satisfies constraint $c' = (\{v_1,v_2,v_4\},\{010,110,011\})$ because $(c \bot A) = \{110\}$, but not for $B=\{0011\}$ because $(c \bot A) = \emptyset$.

Definition 4: A pair $c = (A,B)$ and $c' = (A',B')$ of constraints holds the relation $c \leq c'$, iff $A' \subseteq A$ and the projection of c to A' is smaller than or equal to B, i.e., $c \bot A' \subseteq B'$. For example, for the constraints $c = (\{v_1,v_2,v_4\}, \{000,010,101\})$ and $c' = (\{v_1,v_4\},\{00,10,11\})$ it holds $c \leq c'$, because $(c \bot A') = \{00,11\}$.

Utilizing these relations between constraints, in the following sections constraint representation and satisfaction for IP retrieval is modelled.

8.3.2 Constraint representation and satisfaction

We remember that $\underline{C}$ denotes the set of all possible constraints concerning a set of attributes. Then for each subset C, $C \subseteq \underline{C}$, triple (V,X,C) defines a constraint satisfaction problem (CSP cf. [Meye96]) where X characterizes the problem space and $V=\{v_1,v_2,...,v_r\}$ the set of its components (i.e. IP attributes). Both, set C of constraints as well as CSP (V,X,C) is named unary (or binary) if for each constraint $c = (A,B)$ of C it holds $|A| = 1$ (or $|A| \leq 2$) where $|A|$ specifies the number of attributes of set A.

There is a unique mapping of a binary set C into a symmetric matrix $M = mat(C)$ of r rows. Each element m_{ij} of M corresponds to a binary constraint $(\{v_i,v_j\},m_{ij})$ for $\{v_i,v_j\} \subseteq V$.

The related matrix $M = (m_{ij})$ of a set C of binary constraints $c = (A,B)$ is defined as follows:

$$m_{ij} = \begin{cases} B & \text{if } i \neq j \text{ and there is in C a constraint } c = (\{v_i,v_j\},B) \\ X_i \times X_j & \text{if } i \neq j \text{ and there is no constraint } c = (A,B) \text{ in C with } A = \{v_i,v_j\} \\ B, & \text{if } i = j \text{ and there is in C a constraint } c = (\{v_i\},B) \\ X_i & \text{if } i = j \text{ and there is no constraint } c = (A,B) \text{ in C with } A = \{v_i\} \end{cases}$$

Conversely, matrix $M = (m_{ij})$ corresponds to a set C of constraints $c = (\{v_i,v_j\},m_{ij})$. Additionally, set C can be reduced by deleting such constraints $c = (\{v_i,v_j\},m_{ij})$ for which $m_{ij} = X_i \times X_j$ because it is no restriction. For comparing pairs of matrices we use the relation $\leq$ that is defined by the condition that for the sets of corresponding elements relation $\subseteq$ holds.

Transferring the operations *projection* and *join* defined e.g. in [Heue97] to constraints, an adapted multiplication of matrices can be defined.

The product $M = M' \cdot M''$ of two matrices $M' = (m'_{ij})$ und $M'' = (m''_{ij})$ is a matrix $M = (m_{ij})$ given by

$$m_{ij} = \cap_k (((\{v_i,v_k\},m'_{ik}) \oplus (\{v_k,v_j\},m''_{kj})) \perp \{v_i,v_j\}).$$

Equation $M = M' \cdot M''$ implies the relations $M \leq M'$ and $M \leq M''$.

Moreover, for checking the correctness of a binary query Q, the so-called unary part $M' = unary(M)$ of its matrix representation M is introduced by

$$M' = U^2$$

where the symmetrical matrix $U = (u_{ij})$ is defined by

$$u_{ij} = \begin{cases} m_{ij}, & \text{if } i = j \\ \emptyset, & \text{if } i \neq j \end{cases}.$$

In the general case of a set Q of constraints, *problem reduction* is performed by iterative joining $\oplus(c \mid c \in Q)$ of its elements and results in an empty or unary constraint c'. But for a binary set Q of constraints, it can be shown that *problem reduction* is performed by iterative multiplication of matrix $M = mat(Q)$

$$M'' = M^{r+1} = M^r \leq M^{r-1} \leq \ldots \leq M^2 \leq M$$

with the following results (Figure 8.2):

(I) if there is one element (i,j) in matrix M'' with $m_{ij}'' = \emptyset$ the reduction of query Q is empty, i.e., the query contains contradicting constraints and consequently the user has to correct set Q,

(II) if the binary constraints of Q are not a subset of Cartesian product of its unary constraints, i.e. the inequality $M' \leq M$ is violated, query Q is incomplete. In this case, query Q cannot be decided, and therefore, the user has to correct or to reduce it by backtracking, and

(III) the reduction of query Q corresponds to a unary constraint c'.

IP RETRIEVAL BY SOLVING CONSTRAINT PROBLEMS

Input for *problem solving* are for each IP class of the library a unary constraint c characterizing its properties relevant for retrieval and a binary constraint c'=*con*(M') which is the result of problem reduction for query Q.

For this problem we generalize the *constrained satisfaction decision problem* (CSDP [Meye96]) and its solution can yield one of the following results:

(IV) IP class c does not satisfy query Q and

(V) there is a subset X of IPs in IP class c which satisfies the reduced constraint c' of query Q (Figure 8.2).

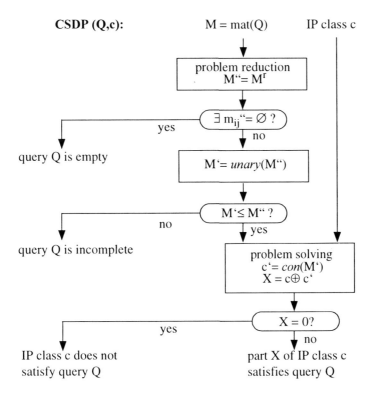

Figure 8.2 Scheme for deciding a generalized CSDP (Q,c)

8.3.3 Retrieval by constraint satisfaction

In order to do IP retrieval, we need a library L of IPs (x,y) and, in the presence of parameterized IP, classes (X,Y) which characterize generalized cases in the problem-solution space. In a generalized case (X,Y) component Y describes the parametrized IP specification, e.g. a synthesized soft core, and component X the set of all vectors of attribute values significant for retrieval. The aim is to search at least one

IP which meets a given set of user-specified constraints of query Q for which it is supposed that its reduction corresponds to a non empty subspace in $\underline{X}$.

For retrieval, problem reduction and solving the generalized CSDP are essential subtasks. When a query Q is incorrect, Q may contain a subset Q' of constraints which meet c. Hence, from constraint set Q a subset Q' is to be selected for backtracking and in the next iteration step the resulting query Q' is to be decided.

After deciding the CSDP (Q,c) for all IP classes c the result of retrieval are classes X of the IP solution space which satisfy the user-specified query Q. Then the next task is to assign by mapping g_2 (Figure 8.1) to each class X the corresponding class of IPs.

8.4 CONCEPT OF AN ADAPTED DATABASE MANAGEMENT

For implementing the reuse approach described in the previous section we developed a concept of a relational DBM-system *(Data-Base-Management)* on the one hand for archiving IP specifications and on the other hand for the presentation of relation attributes for IP retrieval.

8.4.1 System Architecture

The proposed DBM-system enables us to utilize IPs of different structures and complexities (e.g. draft data, test records, test environment, documentation, script for synthesis and simulation) [Rafi98].

For handling data and relations between them we use a flexible and functional extendable system architecture of three layers (Figure 8.3). The first layer, the so-called *User-Host,* is the nearest to the user. It is realized by a Java-applet, which runs as a graphic interface in a Java compatible web-browser. This interface receives the user data inputs and minimizes them concerning computing performance (make structure of the interface elements, character string processing). In this model, several Java-applets can be executed at the same time, i.e., several users can manipulate the same database. The communication with the next layer *(Server-Host)* is based on RMI-functions *(Remote Method Invocation).* Therefore it is possible to develop network applications in a very abstract way.

The second layer is the *Server-Host.* It runs completely on the database processor, and there it executes calculations of independent modular functions (search, data conversion, etc.). The advantage of this layer is, that by subjoin and removing of modules the behavior of the database application can be easily changed. The main task of this layer is the transformation of Java-SQL-requests into database specific SQL-requests *(Structured Query Language).*

Consequently, for adapting or changing of the IP-administration-system for an other database the corresponding SQL-instructions have to be changed. The communication with the *Data-Host*-layer is performed on the base of JDBC-functions

(Java DataBase Connectivity) by means of which the database can be updated and inquired. Utilizing API *(Application Programming Interface)* the communication between Java-programs and the database server is organized by means of SQL instructions. On the right part of Figure 8.3 the structure of the Server-Host is shown in more detail. Various user-applets can be linked in the already described mechanism and use the available services. These services are different functional units, e.g. used for transformation of SQL-queries and specific search functions. By adding new functional units further functionality can be integrated in the system. The lower layer is the *Data-Host*-layer which represents the data model for IP-administration, described in the following section.

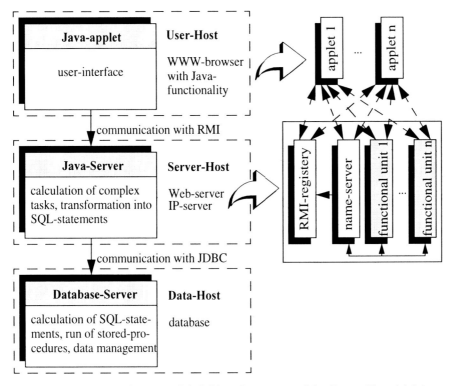

Figure 8.3 Three-layer-model (left) and structure of the Server-Host (right)

8.4.2 Database structure

The *Data-Host* represents the database and its data model for the IP management. The advantage of this layer model is that database system and data models on the lowest layer can be exchanged without the necessity to adapt the upper layer. In Figure 8.4 the data model is illustrated by an ER-diagram *(Entity-Relationship)*.

114 VIRTUAL COMPONENTS DESIGN AND REUSE

Here, an entity is described by a rectangle and represents an object or case which has to be implemented as a physical database object. Relations are given by rhombuses and represent the schematic coherency between the entities. The inscription at the lines between entity and relation characterize the cardinality, i.e. the possible range, described by <n,m>. Sign „*" means, that the cardinality is open on the top. In the data model, several views are assigned to an IP. Such a view can be a VHDL-Description, a Design Ware Component or a simple VHDL-package-function. The identifier of different views are archived in a table. If a user needs a special view that is not archived in the table, then it can be generated. In order to find an already archived view, only the instance of this view is to be specified. Also an IP documentation can be assigned to a view. By this data structure implementation it is possible to manage different structured IP data in one data model.

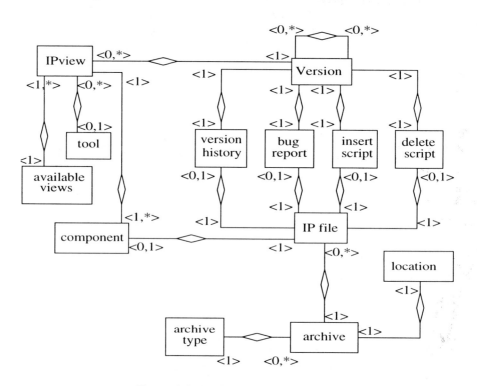

Figure 8.4 Entity-Relationship-Model

Figure 8.5 shows a part of the data structure of a 8051-micro controller IP from the FhG-IIS-Erlangen [Desi99]. The views needed for this data structure could be e.g. documentation and control data, DesignWare, libraries, simulation procedure, synthesis flow and test bench. All these views are instances of referenced tables and could be used for further IPs. Thereby, it is possible to file different structured IPs in the database. Each view consists of an amount of IP data (component, IP). These

IP RETRIEVAL BY SOLVING CONSTRAINT PROBLEMS 115

data are assigned to archives, e.g. characterized by *.*tar* or also as *.*gz*. One IP version belongs to each IP, which is filed in an archive of its own.

For IP retrieval additional attributes are needed to specify it. Especially, the relations between these attributes must be stored in the database by a set of referenced tables. These tables are dynamically generated when the IP characteristics are inserted and are catenated with each other building a graph structure. Each table of a fixed attribute stores the relation between its predecessors and its successor, i.e., the table contains a subset of the Cartesian product of these two sets.

The internal description of the relations is a pointer structure, which is saved in the nodes of the graph. The sheets contain basic data, which cannot be branched further (e.g. bitwidth). This data organization has the advantage, that in comparison with the traditional description tree, it is not necessary to flow every possible path for information, consequently the process speed is higher.

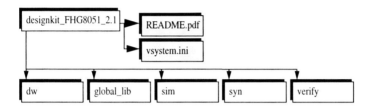

Figure 8.5 Part of the structure of an 8051-IP (FhG-IIS-Erlangen)

8.5 SUMMARY

Our contribution is focused on the retrieval of IPs from a given library of synthesized, verified, well documented designs characterized by a set of attributes. As a new approach we employ case-based reasoning and constraint satisfaction techniques for IP retrieval [Oehl98]. Modeling sets of binary constraints by matrices map the main steps of IP retrieval into matrix multiplications.

Additionally, the representation of IPs by constraints is also the basis for *similarity based* retrieval of parameterized IPs. This is the key to problem solution in case-based reasoning systems as they are being developed by artificial intelligence research [Berg99a]. Similarity measures for IPs represented by constraints are further elaborated in [Berg99b].

Based on a modular three layer system architecture we propose a database structure for implementing this retrieval approach, which is characterized by a two view representation of an IP in the database, namely the data and relation view. All information specifying the IP, like keywords, properties and designs [Reut99], are stored in the data view, whereas the relations between the IP attributes, necessary for retrieval, belong to the relation view.

8.6 EXAMPLE

The introduced framework will be illustrated by a simple example, namely the German-map coloring problem [Meye96]. The task is to assign the countries of the map of Germany using only three colors: red (r), green (g), and blue (b), but considering the condition that neighboring countries have different colors. For searching a specific correct coloring a set of constraints can be specified.

There is the following relation between the coloring task and the IP retrieval:

- Set $V = \{v_1, v_2\}$ of countries corresponds to the set of IP attributes. For simplification we suppose that Germany consists of two countries v_1 and v_2 only.
- Set $\{r,g,b\}$ of colors characterizes the domains $\underline{X}_1 = \underline{X}_2 = \{r,g,b\}$ of the IP attributes.
- Set $\underline{X} = \underline{X}_1 \times \underline{X}_2 = \{(r,r),(r,g),(r,b),(g,r),(g,g),(g,b),(b,r),(b,g),(b,b)\}$ of all possible colorings of two countries represents the constraints for characterizing IPs. Set $\underline{X}$ corresponds to the problem space.
- Set $\underline{Y} = \{y_1, y_2,...,y_9\}$ shall be the set of IP specifications, especially their names. Set $\underline{Y}$ corresponds to the solution space.
- The Cartesian product of $\underline{X} \times \underline{Y} = \{((r,r),y_1), ((r,g),y_2), ((r,b),y_3), ((g,r),y_4), ((g,g),y_5), ((g,b),y_6),((b,r),y_7), ((b,g),y_8), ((b,b),y_9)\}$ describes the mapping f from the set of IP specifications $\underline{Y}$ into the set of IP constraints $\underline{X}$ (see Figure 8.1).
- Set $\underline{X} \times \underline{Y}$ corresponds to the problem-solution space.
- A subset L of set $\underline{X} \times \underline{Y}$ can be interpreted as an IP library. Each IP of this library is given by a pair (x,y) where y is the specification respectively the name of the IP and x the constraint characterizing it.
  May $L = \{((r,g),y_2),((r,b),y_3),((g,r),y_4),((g,b),y_6),((b,r),y_7),((b,g),y_8)\}$ be an IP library, then the complete class of IPs can be characterized by the unary constraint $c = (\{v_1,v_2\},\{(r,g),(r,b),(g,r),(g,b),(b,r),(b,g)\})$.

In the case of two countries v_1 and v_2, a query $Q = (c_1, c_2, c_{12})$ for searching a correct coloring can consist of two unary constraints c_1 and c_2, and one binary constraint c_{12}. Constraint c_1 and c_2 may describe the preferred colors of the countries v_1 and v_2, respectively. Moreover, there is a restriction c_{12} of the government for coloring this two countries.

In the following we discuss three cases of query $Q = \{c_1, c_2, c_{12}\}$ for finding a correct coloring of the two countries v_1 and v_2.

First case: (An example of an empty query.)

Applying the constraints

$c_1 = (\{v_1\},\{r\})$, $c_2 = (\{v_2\},\{g,b\})$, and $c_{12} = (\{v_1,v_2\},\{(g,r),(b,b),(b,r)\})$,

IP RETRIEVAL BY SOLVING CONSTRAINT PROBLEMS 117

the matrix representation of query $Q = \{c_1, c_2, c_{12}\}$ is given by:

$$M = \begin{bmatrix} \{r\} & \{(g,r),(b,b),(b,r)\} \\ \{(g,r),(b,b),(b,r)\} & \{g,b\} \end{bmatrix}.$$

Problem reduction for M is performed by matrix multiplication:

$$M^2 = \begin{bmatrix} \emptyset & \emptyset \\ \emptyset & \{b\} \end{bmatrix} = M'' = 0,$$

and consequently, the first query is empty.

Second case: (An example of an incorrect query.)

Applying the constraints

$c_1 = (\{v_1\},\{r\})$, $c_2 = (\{v_2\},\{g,b\})$, and $c_{12} = (\{v_1,v_2\},\{(r,g),(b,b),(b,r)\})$,
the matrix representation of query $Q = \{c_1, c_2, c_{12}\}$ is given by:

$$M = \begin{bmatrix} \{r\} & \{(r,g),(b,b),(b,r)\} \\ \{(r,g),(b,b),(b,r)\} & \{g,b\} \end{bmatrix}.$$

Problem reduction for M is performed by matrix multiplication:

$$M^2 = \begin{bmatrix} \{r\} & \{(r,g)\} \\ \{(r,g)\} & \{g,b\} \end{bmatrix} = M''.$$

Since the problem reduction M" is not empty, the correctness of the reduced query to be checked by comparing it with the unary matrix $M' = unary(M)$ of M. The unary matrix $M' = U^2$ is the square of the matrix

$$U = \begin{bmatrix} \{r\} & \emptyset \\ \emptyset & \{g,b\} \end{bmatrix}$$

specified by the unary constraints $c_1 = (\{v_1\},\{r\})$ and $c_2 = (\{v_2\},\{g,b\})$:

$$M' = \begin{bmatrix} \{r\} & \emptyset \\ \emptyset & \{g,b\} \end{bmatrix}^2 = \begin{bmatrix} \{r\} & \{(r,g),(r,b)\} \\ \{(r,g),(r,b)\} & \{g,b\} \end{bmatrix}.$$

Since condition $M' \leq M''$ is violated, query Q is incorrect, and therefore it is to be corrected, e.g. by reducing the unary constraint c_2 (compare the third case).

Third case: (An example of a correct query.)
Applying the constraints
$c_1 = (\{v_1\},\{r\})$, $c_2 = (\{v_2\},\{g\})$, and $c_{12} = (\{v_1,v_2\}, \{(r,g),(b,b),(b,r)\})$,
the matrix representation of query $Q = \{c_1, c_2, c_{12}\}$ is given by:

$$M = \begin{bmatrix} \{r\} & \{(r,g),(b,b),(b,r)\} \\ \{(r,g),(b,b),(b,r)\} & \{g\} \end{bmatrix}$$

Problem reduction for M is performed by matrix multiplication:

$$M^2 = \begin{bmatrix} \{r\} & \{(r,g)\} \\ \{(r,g)\} & \{g\} \end{bmatrix} = M".$$

Since the problem reduction M" is not empty, the correctness of the reduced query is to be checked by comparing it with the unary matrix M' = *unary*(M) of M.

The unary matrix M'= U^2 is the square of the matrix

$$U = \begin{bmatrix} \{r\} & \emptyset \\ \emptyset & \{g\} \end{bmatrix}$$

specified by the unary constraints $c_1 = (\{v_1\},\{r\})$ and $c_2 = (\{v_2\},\{g\})$:

$$M' = \begin{bmatrix} \{r\} & \emptyset \\ \emptyset & \{g\} \end{bmatrix}^2 = \begin{bmatrix} \{r\} & \{(r,g)\} \\ \{(r,g)\} & \{g\} \end{bmatrix}.$$

Since condition M'≤ M" is satisfied, query Q is correct.

Moreover, the reduction of the given query Q is represented by the binary constraint c'= *con*(M) = $(\{v_1,v_2\},\{(r,g)\})$.

The second task is problem solving, i.e. it is to retrieve at least one element of the library L = $\{((r,g),y_2),((r,b),y_3),(_4(g,r),y),((g,b),y_6),((b,r),y_7),((b,g),y_8)\}$ which satisfies constraint c'= $(\{v_1,v_2\},\{(r,g)\})$. In our example IP $((r,g),y_2)$ is the solution, i.e., a correct coloring of the two countries is color red for country v_1 and color green for country v_2.

With respect to Figure 8.2 by mapping g_1, constraint satisfaction results in the reduced query X = g_1(Q) = $(\{v_1,v_2\},\{(r,g)\})$. After that, by mapping g_2, the IP set Y = g_2(X) = $\{((r,g),y_2\ ))\}$ of the library L is assigned to the reduced query X.

9 CRYPTOGRAPHIC REUSE LIBRARY

Andreas Schubert, Ralf Jährig and Walter Anheier

University of Bremen
Institut für Theoretische Elektrotechnik und Mikroelektronik (ITEM)
Bremen, Germany

9.1 INTRODUCTION

In recent years the importance of security in information technology has increased drastically. The main reason is the obvious demand for security software and hardware in growing applications like mobile telephones, set top boxes, wireless networks, smartcards and other distribution systems for entertainment and information. In addition to higher processing speed security hardware offers higher physical manipulation protection than software solutions. In connection with the growing core-based system design methods a cryptographic reuse library based on VHDL is presented in this contribution.

The core-based design methodology is an essential approach for increasing the productivity in system design. It is characterized by structured design methods and reuse of existent components and subsystems. Reuse libraries with configurable virtual components (VCs) are the basis for core-based design methods [Jerr97]. In this work the developed reuse library contains virtual components with cryptographic functions, i.e., algorithms to ensure confidentiality, authenticity, integrity and/or commitment of data and messages.

In a first development step a library basis is created, which consists of optimized cryptographic virtual components (VCs) based on various modern symmetric block ciphers (e.g. SAFER K-128, 3WAY or RC5 [Schn94]) with different security, performance and implementation cost features. By using the modes of operation according to standard ISO/IEC 10116 sym. block ciphers cover a broad range of security tasks [ISO91]. Further modern sym. block ciphers can be integrated into the library (cf. developments in the AES initiative of US NIST for example

[Nist99]). The register-transfer level models of the cryptographic reuse library are based on the standardized hardware description language VHDL. Each of the cryptographic virtual components of the library consists hierarchically of three submodules.

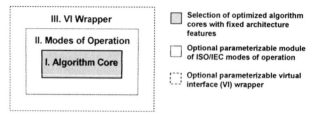

Figure 9.1 Cryptographic virtual component (CVC)

9.2 LIBRARY CONCEPT

The elements of this library are so-called soft VCs (soft cores), i.e., to a large extent technology-independent (assuming availability of memory macro cells in target libraries) and synthesizable models of implemented symmetric block ciphers at register-transfer level (RTL). These virtual components belong to the group of custom functional blocks. Sym. block ciphers are roughly comparable with the class of typical DSP algorithms like FFT or digital filters with regard to functional complexity. In contrast to these DSP functions, computations in sym. block ciphers are based on non-scalable operations with integers. This means, in case of mapping a sym. block cipher to a corresponding reusable virtual component a parameterization of data path widths and accuracy is not possible.

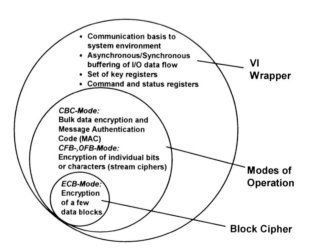

Figure 9.2 Increasing application flexibility by optional wrapper modules

In order to reduce the disadvantage of non-scalability, the library provides several optimized, non-parameterizable core architectures of various symmetric block ciphers with different algorithm parameters, performance features, hardware costs and security properties. Optionally, the algorithm core can be embedded hierarchically in a module to realize the ISO/IEC modes of operation and a virtual interface wrapper (see Figure 9.1). The two outer optional modules extend the application range of the algorithm cores by a number of additional functions as illustrated in Figure 9.2.

Hardware configuration	Software configuration
Selection of algorithms/architectures with specific features (data path width, pipelining degree etc.)	Selection of modes of operation: e.g. ISO/IEC modes of operation
Parameterization of data path widths	Selection of functional registers
Parameterization of number of functional registers	
Parameterization of width and depth of registers and memories	

Table 9.1 Typical hardware and software configuration areas for CVCs

Generally, the reuse of the cryptographic virtual components is enabled by different forms of configuration. An important aspect is the identification of configurable structures and functions and their assignment to hardware- or software-based configuration measures. Typical configuration areas for the CVCs are given in Table 9.1.

The configuration of the CVCs is carried out from inside to outside as shown in Figure 9.3, i.e., from the algorithm core (I, lower hierarchy level) over the module of the ISO/IEC modes of operation (II, middle hierarchy level) to the virtual interface wrapper (III, upper hierarchy level).

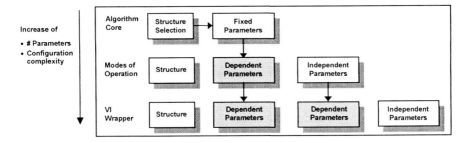

Figure 9.3 Procedure of CVC hardware configuration

A part of the parameters of the hierarchy levels II and III are dependent parameters. They have to be adapted to the corresponding parameters at the next lower hierarchy level. Note that together with the application flexibility the configuration complexity increases from the lower to the upper hierarchy level.

Reuse methodology

The two essential methods to create degrees of freedom in component models and to increase the reusability of the components are parameterization of signal and port widths and structure selection.

The applied modeling style with regard to these features is orientated by the VHDL-based methodology described in [Reut97]. It is suitable for both synthesis and simulation:

- The parameterization is enabled by a generic interface (VHDL construct "generic"). For hierarchical circuit structures it allows:
    - Controlling the bit widths of ports and signals.
    - Generic parameter passing to components for selection of internal submodules (by means of VHDL construct "generate") and configuration of controllers.
- The VHDL construct "generate" is the basis for structure selection: Choice of structures and functions (dependent on generic parameters).

Constant factors for setting the parameters are written in appropriate packages. The VHDL construct "configuration" is not used for configuration tasks, since it is not completely supported by synthesis.

Algorithm	Source	n (Bit)	l (Bit)	#Encryption rounds	Operation width (Bit)	On-the-fly subkey comp.
RC5-32/12/16	R.L. Rivest, 1994	64	128	12	32	no
SAFER K-128	J.L. Massey, 1994	64	128	10 (- 12)	8	no
3WAY	J. Daemen, 1993	96	96	11 (eff. 12)	32	yes

Table 9.2 Algorithm properties of the used symmetric block ciphers

9.3 CVC STRUCTURE

9.3.1 Algorithm core

The available algorithm cores are based on different modern symmetric block encryption algorithms. A block cipher converts a plain text block $P = (p_1, p_2, ..., p_n)$ of a fixed bit width into a cipher text block $C = (c_1, c_2, ..., c_n)$ of the same size: i.e., $C = eK(P)$ with encryption function eK and key $K = (k_1, k_2, ..., k_l)$ and vice versa $P = dK(C)$ with decryption function dK. Together with the selection of the algorithm (eK and dK) the corresponding parameters data block width n and key length l are determined.

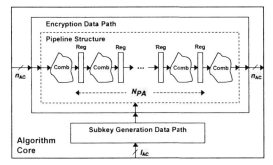

Figure 9.4 Simplified VLSI architecture of the algorithm core

An overview of the used algorithms and further important properties is shown in Table 9.2. The degree of cryptographic security is characterized by algorithm quality, key length, data block width and number of encryption rounds. The algorithms are mapped to optimized and synthesizable VLSI architectures at register-transfer level (soft cores). The abstract block diagram of a generalized VLSI architecture for sym. block ciphers is shown in Figure 9.4.

Algorithm	Architecture	n_{AC} (Bit)	l_{AC} (Bit)	N_{PA}	Latency L_{Enc} (clock cycles)	Latency L_{KeyExp} (clock cycles)
RC5-32/12/16	Serial crypto unit	64	128	1	75	494
SAFER K-128	1 Round impl.	64	128	6	61 (73[a])	24 (28[a])
3WAY, I. Arch.	1 Round impl.	96	96	1	12	-
3WAY, II. Arch.	4 Round impl.	96	96	4	12	-

a 12 encryption rounds

Table 9.3 Architecture features of the algorithm soft cores

An overview of the different technology-independent qualities of the developed soft cores can be found in Table 9.3. Some sym. block ciphers (e.g. 3WAY) allow a computation of the subkeys simultaneously with the encryption process (on-the-fly subkey calculation).

In some cases pipelining techniques are applied in the data paths of the VLSI architectures (N_{PA} = pipelining degree). To determine the performance of the developed RTL architectures they are mapped to a reference library (0.7 μm CMOS process). The results are firm cores (gate-level netlists). The technical data of the firm cores are listed in Table 1.4.

Algorithm	App. area (mm²)	Max. clock frequency (MHz)	Max. data throughput (MBit/s)
RC5-32/12/16	12	30	25.6
SAFER K-128	23	40	251.8 (210.4[a])
3WAY, I. Architecture	9	55	440
3WAY, II. Architecture	20	50	1600

a 12 encryption rounds

Table 9.4 Performance and cost features of the firm cores (0.7 μm CMOS)

9.3.2 Module of the modes of operation

The block cipher module as the core of the cryptographic virtual component is integrated in the adaptive module of the modes of operation (see Figure 9.5, left). The modes of operation according to ISO/IEC international standard 10116 improve the cryptographic security of n-bit block ciphers and extend their application range (e.g. bulk data encryption, stream encryption, authentication etc.). The modes of operation are described by the following encryption and decryption procedures. Consider a sequence of q plain or cipher text blocks. For $i = 1, 2, ..., q$:

Encryption   *Decryption*

n-bit ECB $e: C_i = eK(P_i)$ n-bit ECB $d: P_i = dK(C_i)$

n-bit CBC $e: C_i = eK(P_i \oplus C_{i-1})$, $C_0 = SV$ n-bit CBC $d: P_i = dK(C_i) \oplus C_{i-1}$, $C_0 = SV$

j-bit CFB $e: Y_i = eK(X_i)$ j-bit CFB $d: Y_i = eK(X_i)$
$\qquad C_i = P_i \oplus SL_j(Y_i)$ $\qquad P_i = C_i \oplus SL_j(Y_i)$
$\qquad X_{i+1} = S_j(X_i|C_i), X_1 = SV$ $\qquad X_{i+1} = S_j(X_i|C_i), X_1 = SV$

j-bit OFB $e: Y_i = eK(X_i)$ j-bit OFB $d: Y_i = eK(X_i)$
$\qquad C_i = P_i \oplus SL_j(Y_i)$ $\qquad P_i = C_i \oplus SL_j(Y_i)$
$\qquad X_{i+1} = Y_i, X_1 = SV$ $\qquad X_{i+1} = Y_i, X_1 = SV$

Note that $SL_j(Y) = (y_1, ..., y_j)$ and $S_j(X|C) = (x_{j+1}, x_{j+2}, ..., x_n, c_1, c_2, ..., c_j)$ for $j \leq n$.

Data block width n, key length l and the reduced data block width j are parameters in the above equations. Figure 9.5 (left) and Table 9.5 show the hardware and software configuration possibilities for the module of modes of operation.

Parameter	Values
Conversion direction of hierarchy level II	Encryption or Decryption
Modes of operation	ECB, CBC[a] (MAC),CFB or OFB or Bypass
Reduced data block widths in CFB and OFB mode (bit)	j_1, j_2, j_3 or j_4
Number of data blocks per conversion	1 or N_{DM}

a optional: SV encryption

Table 9.5 Software configuration of the module of modes of operation

9.3.3 Virtual interface wrapper

The virtual interface (VI) wrapper depicted in Figure 9.6 is a flexible and generalized basis for different types of communication. This concerns both physical connections and protocols.

The VI wrapper is also characterized by its high-performance architecture. By means of two asynchronous input and output FIFO memories the I/O data flow can be buffered. Thus, encryption/decryption and data transmission can be done in parallel.

The FIFO memories have parameterizable width and depth. In case of the asynchronous FIFOs a selection between a technology-dependent Dual-Port RAM architecture and a technology-independent so-called ping-pong architecture based on flip flops is possible. Additionally, there are two parameterizable modules for the data width conversion between external buses and the FIFOs and two parameterizable modules to reduce the actual block width in CFB or OFB mode.

In the VI wrapper, parameterizable commands and status registers can be found. Besides parameterization of the number and width of these registers, the mapping of register bits to the command and status signals is also parameterizable. The addresses of registers and the positions of bits in the registers can be defined by the user.

Finally, the wrapper has a variable number of parameterizable write-only key registers. For example, several key registers are needed when the cryptographic VC has to handle different secure communication channels simultaneously.

The key registers and the command registers can be logically separated from each other by different parameterizable address areas. The direct key availability in the CVC and the write-only property is a fast and secure solution for this kind of applications ([Ande88] [Bonn93]). Table 9.6 shows the software configuration possibilities of the VI wrapper.

126 VIRTUAL COMPONENTS DESIGN AND REUSE

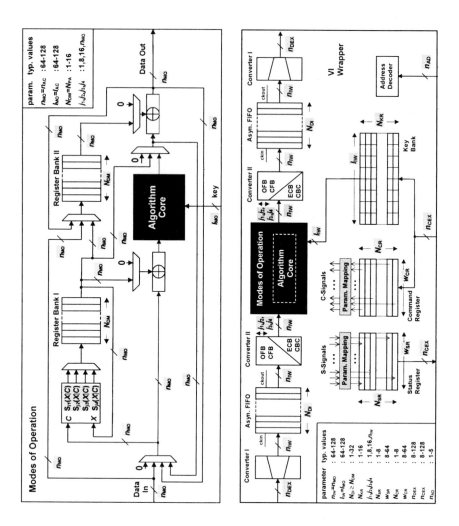

Figure 9.5 Hardware configuration of the module (left)
Figure 9.6 Hardware configuration of the VI wrapper (right)

Parameter	Values
Reduced data block widths in CFB and OFB mode (bit)	j_1, j_2, j_3 or j_4
No. of the key register to be used	1, 2, ... or N_{KR}
No. of the status or command register to be used	1, 2, ... or N_{SR} / 1, 2, ... or N_{CR}

Table 9.6 Software configuration of the VI wrapper

9.4 CVC INTERFACES

9.4.1 Internal interfaces

There are two internal interfaces between the algorithm core, the module of modes of operation and the VI wrapper (see Figure 9.7). They consist of scalable buses for data and keys (type A) as well as command and status signals of fixed bit width. The data and control communication through the interfaces takes place synchronously to the internal clock. The command and status signals of the two internal interfaces can be classified as follows:

- Signals (type B) for software configuration (e.g. modes of operation),
- Signals (type C) to indicate start and end of processing tasks (e.g. encryption) and
- General signals (type D) like clock or reset.

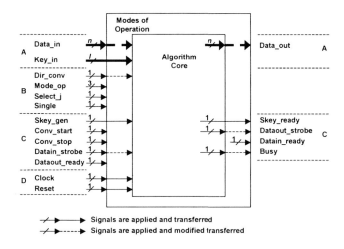

Figure 9.7 Internal interfaces

Similar to the implementation of the control tasks (FSM) the control signals are organized in such a way that requirements like modularity or hierarchy are met. Thus, the control signals of the two internal interfaces are formed in such a way that the respective module can be embedded in the next higher hierarchy level or operated separately. The hierarchical control of the processes like subkey generation or encryption takes place independently of the specific hardware configuration (e.g. independent of latencies of the used algorithm core) as shown in Figure 9.8.

128   VIRTUAL COMPONENTS DESIGN AND REUSE

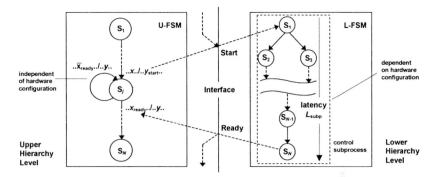

Figure 9.8   Procedure for connection of FSMs at different hierarchy levels

9.4.2 Virtual interface

The objective of the development of the virtual interface (VI) is that the cryptographic virtual component is compatible with or easily adaptable to as many as possible different communication structures. Therefore, following design guidelines for the development of the virtual interface are pursued [Prid95]:

- The aim is to dtermine an interface encapsulation as a general basis for a communication structure but not a detailed description of a physical interface. The interface wrapper should have as little as possible functions of a specific interface.

- The virtual interface should have a simple, flexible and technology-independent structure, which is focused to the particular interface aspects of cryptographic virtual components.

- Different communication paradigms (on-chip bus systems, board-level bus systems or specific point-to-point connections - e.g. direct connection with network interfaces) have to be considered.

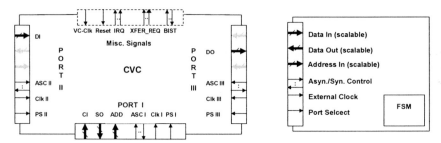

Figure 9.9   (a) Virtual interface of the CVC (b) Reusable I/O module

In our opinion the approach depicted in Figure 9.9(a) is close to the objectives listed above. It is characterized by the following features:
- The CVC always behaves as a slave.
- All buses are scalable and unidirectional (technology-independence by avoiding tristate logic).
- The VI has three ports: data input port (II), data output port (III) and port I with command/key input bus, address input bus and status output bus.
- The ports have basic control signals for asynchronous/synchronous communication.
- The VI has additional control signals for interrupt, transfer request and BIST.

The control signals are active low (not marked in the figures for reasons of simpler representation). For modeling the three ports, the same general I/O module is used (see Figure 9.9(b)). This module provides two unidirectional data buses with opposite directions for asynchronous or synchronous data transmission and an additional address bus. The buses and the control signals are partially optional. Based on the generalized interface the following communication paradigms for system integration are possible:
- A common bidirectional bus (data, control and key) for connection with an individual bus (e.g. microprocessor bus or on-chip bus).
- An unidirectional input data bus and an unidirectional output data bus (e.g. for point-to-point connection with functional modules like network interfaces) and a bidirectional control bus which is connected with an appropriate bus (e.g. peripheral bus).
- Two separate bidirectional data buses for plain and cipher text and a separate bidirectional control bus.
- Two unidirectional input buses (data and command) and two unidirectional output buses (data and status) for connection with bus structures based on multiplexer techniques (e.g. technology-independent on-chip bus systems).

9.5 CONCLUSIONS

The cryptographic reuse library presented in this work provides a number of optimized cryptographic virtual components (VCs). They are based on various modern sym. block ciphers with different security, performance and hardware cost features. The cryptographic virtual component consist of an algorithm core which is embedded in an adaptive and parameterizable module to realize the modes of operation. This allows the flexible use of the cryptographic VCs in a large range of security applications. The additional virtual interface wrapper is a flexible I/O frame, which is a good basis for a simple integration into different system environments. The module of modes of operation and the virtual interface wrapper are constructed in such a way that further algorithm cores can be easily integrated into the library. In order to increase the capabilities of the library, it is useful to consider cryptographic cores based on asymmetric algorithms (e.g. RSA) for a future library extension. Asymmetric algorithms have distinct advantages regarding key management and digital signature.

10 A VHDL REUSE COMPONENT MODEL FOR MIXED ABSTRACTION LEVEL SIMULATION AND BEHAVIORAL SYNTHESIS

Cordula Hansen, Oliver Bringmann
and Wolfgang Rosenstiel

FZI Karlsruhe
Department Microelectronics System Design
Karlsruhe, Germany

10.1 INTRODUCTION

Due to the increasing complexity of digital systems, it is often desirable to start the design at higher levels of abstraction, e.g. at the algorithmic level. The necessary transformations are then performed by commercial or scientific high-level synthesis systems. In complex system design, the integration of user defined RT components (IP blocks) in the algorithmic specification is getting more and more important for the following reasons. First, several RT components appropriate for reuse may already exist. Second, the re-implementation of VHDL models emulating this behavior at the algorithmic level is expensive and time-consuming. Third, some functional and timing behavior can only be implemented at the RT level, e.g. interrupt handling, and interface components. Finally, several synthesis, simulation, and test environments which can be used for descriptions on different abstraction levels are already available. Therefore, this contribution addresses the problem of mixed abstraction level specifications for simulation and behavioral synthesis to allow the reuse of existing RT components.

For this, the VHDL standard [IEEE93] without any extensions is used and the usual simulation and synthesis systems can be applied. The communication between algorithmic descriptions and VHDL components at the same or at lower levels is executed using VHDL procedures. To reduce the design time required for the insertion of these procedures in the algorithmic specification, a preprocessor has been developed. The preprocessor allows the procedures to be applied without any extensive declarations of the corresponding RT components. The implementation of procedures emulating the component behavior at the algorithmic level is also possible.

The different timing aspects at algorithmic, RT, or lower levels are encapsulated in the provided procedures. As a result, the algorithmic specification does not have to contain any timing aspects to communicate with components at lower abstraction levels. Finally, an appropriate stimuli set can easily be included testing the integrated VHDL component or the procedure together with the component.

10.1.1 Related Work

A closer investigation of existing approaches shows that two different methods can be identified. First, there exist several object-oriented (OO) approaches extending VHDL with the corresponding OO language constructs (OO-VHDL) ([Ashe98], [Jerr97], [Rade97], [Schu95], [Swam95]). Using one of these approaches, a mixed abstraction level simulation and synthesis can easily be executed. However, an extension of the standardized language VHDL is necessary. Consequently, commercial simulators and synthesis systems, as well as most of the scientific tools, can not be applied. Second, in [Kiss95] a structured design methodology is presented. In this approach, procedures are also applied implementing the communication to the corresponding RT components. However, each time such a procedure is used the implementation of the procedure as well as the declaration of the corresponding RT component must be manually inserted in the specification. Furthermore, the synthesis process has no optimization potential for RT components using the same kind of procedures.

This chapter is organized as follows: Section 10.2 and Section 10.3 give a general overview of our VHDL reuse component model for mixed abstraction level simulation and synthesis. In Section 10.4, the creation of a reuse library and the integration into our high-level synthesis system CADDY-II are described. Section 10.5 and Section 10.6 present further synthesis and simulation aspects concerning our approach. This chapter concludes with a summary in Section 10.7.

10.2 A VHDL REUSE COMPONENT MODEL FOR SIMULATION AND SYNTHESIS

High-level synthesis systems transform an abstract behavioral specification into a structure of RT components and a finite state machine. Therefore, the main problem for high-level synthesis with mixed abstraction level simulation is to provide a comfortable environment for integrating RT components into an algorithmic specification. Using the hardware description language VHDL, algorithmic descriptions are specified in a VHDL process. The reuse of RT components is realized by instantiating the needed RT components within the actual specification. However, component instantiation is not allowed inside a VHDL process, using the VHDL standard without any extensions. One solution would be a structural description, but in this case, the timing signals needed at the RT level, e.g. clock and reset signals, must also be used in the algorithmic specification although they are generated during high-level synthesis. To overcome this problem procedures are used to implement the communication with the RT components (*access procedures*). To reduce the design time required for the insertion of these procedures in the algorithmic specification, a preprocessor has been developed. The preprocessor allows the usage of procedures without time-consuming declaration of these procedures and their corresponding RT components.

The use of procedures has several advantages. Procedures can be used in VHDL processes. The RT timing as well as the communication protocols can be encapsulated in the access procedures. Every procedure has a corresponding RT component forming a unit for simulation (communication aspect) and synthesis (correspondence aspect) (Figure 10.1).

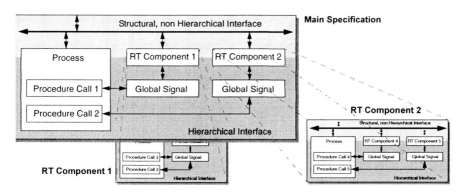

Figure 10.1 Hierarchical specification using access procedures

To implement this unit in VHDL without any language extensions, in our approach, we use the "frame component" concept. Every frame component contains the RT component to be reused and all corresponding access procedures. Hence, the synthesis process can use the information about the relation between access procedures and RT components, expressed by the frame component, for an optimized integration of the RT components. Furthermore, the generated mixed abstraction level VHDL description including the integrated RT components can be simulated using standard VHDL simulators. The frame component is implemented as a usual VHDL component and the entire component library is described in standard VHDL. As a result, a reuse component library can be easily created and used by the presented VHDL library concept.

134   VIRTUAL COMPONENTS DESIGN AND REUSE

10.3 THE FRAME COMPONENT CONCEPT

As mentioned above, the basic unit of our approach is the frame component. The frame component is a standard VHDL component, and contains one user defined RT component and its corresponding procedures. For every RT component, a frame component has to be defined once. Each procedure represents an operation which can be performed by an RT component. The communication between an access procedure and the corresponding component is performed by global signals defined in the declarative part of the frame component. A simple example of a frame component for a gcd algorithm can be seen in Table 10.1. In this example, a frame component *frame_comp_gcd* with ENTITY and ARCHITECTURE specification has been defined. As usual, the ENTITY specification can contain generic parameters and interface signals.

The generic parameters are taken from the ENTITY of the user defined component. Concerning the interface signals, it has to be distinguished between the interface signals of the ARCHITECTURE containing the algorithmic specification (main specification), of the processes representing the algorithmic specification (algorithmic specification), of the RT components, of the frame components, and of the access procedures. An overview of the relationships between these interfaces is given in Figure 10.2.

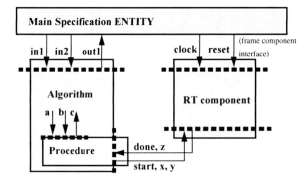

Figure 10.2 Interface relations

The interface of the main specification has to consist of the interface signals of the algorithmic specification and those of the frame component. The interface signals of the frame component correspond to the interface signals of the RT component that are not contained in the parameter list and in the body of the corresponding access procedures, e.g. Table 10.1 (line 6). Thus, all external signals required for the algorithm and for the synthesized RT implementation are contained in the interface of the specification. As a result, the interface will not be changed during synthesis. The implementation of the specification interface can be done manually or by using our VHDL preprocessor which automatically extends the specification interface.

```
(1)     LIBRARY ieee;
(2)     USE ieee.std_logic_1164.ALL; USE ieee.std_logic_arith.ALL;
(3)
(4)     ()ENTITY frame_comp_gcd IS
(5)     () GENERIC( width1 : integer);
(6)        PORT( clock,reset : IN std_logic);
(7)     END frame_comp_gcd;
(8)
(9)     ARCHITECTURE specification OF frame_comp_gcd IS
(10)       COMPONENT gcd_rt_comp
(11)          GENERIC (width1 : integer);
(12)          PORT(clock, reset : IN std_logic;
(13)             start : IN std_logic;
(14)             done : OUT std_logic;
(15)             x,y : IN signed(width1-1 DOWNTO 0);
(16)             z : OUT signed(width1-1 DOWNTO 0)
(17)          );
(18)       END COMPONENT;
(19)       SIGNAL start, done : std_logic;
(20)       SIGNAL x,y,z : signed (width1-1 DOWNTO 0);
(21)    BEGIN
(22)
(23)       gcd1 : gcd_rt_comp
(24)          GENERIC MAP (width1);
(25)          PORT MAP (clock, reset, start, done, x,y,z);
(26)       PROCESS
(27)          -- declarative part
(28)          PROCEDURE gcd(a, b, c : IN signed(width1-1 DOWNTO 0)) IS
(29)          BEGIN
(30)             specification.start <= '1';
(31)             specification.x <= a;
(32)             specification.y <= b;
(33)             WAIT UNTIL specification done'event
(34)                   AND specification.done = '1';
(34)             specification.start <= '0';
(35)             c := specification.z;
(36)             WAIT FOR 1 ns;
(37)          END gcd;
(38)       BEGIN
(39)          -- stimuli set for access procedures/RT component
(40)       END PROCESS;
(41)    END specification;
```

Table 10.1 Frame component for a simple GCD example

In the ARCHITECTURE specification, the user defined RT components are instantiated and the corresponding procedures are implemented. These procedures can be access procedures or emulation procedures. For simulation, the implementation of a frame component for emulation procedures is not absolutely necessary (Section

10.6). The frame component is used here to give the synthesis system the necessary mapping information, and to increase the optimization potential (Section 10.5). In contrast, for access procedures, the frame component is required for simulation as well as for synthesis. An important advantage of this approach is that both kinds of procedures allow the implementation of user defined components at any arbitrary abstraction level supported by VHDL (Figure 10.4).

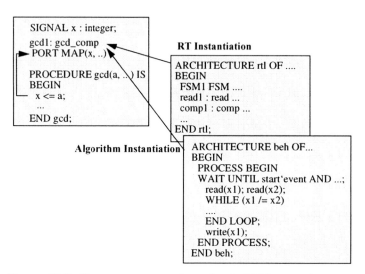

Figure 10.3 Frame component for a simple GCD example

However, to implement an access procedure, the declarative part of the ARCHITECTURE of the frame component has to contain the declaration of the user defined component and all signals required for the communication between the procedure and the component (Table 10.1 (line 10-20)). In the implementation part, the component is instantiated, and the access procedures are implemented in the declarative part of a PROCESS (Table 10.1 (line 27-37)). The implementation part of the PROCESS can particularly be used for specifying simulation vectors in order to test the component including the corresponding procedures (Table 10.1 (line 39)).

10.4 REUSE COMPONENT LIBRARY AND CADDY-II

An important advantage of the frame component concept is that a reusable component library can easily be implemented using the VHDL library concept. For every frame component a corresponding component declaration can be inserted in a VHDL package (Table 10.2). This allows an easy extension of the component library.

```
(1)  PACKAGE reuse_comp_library IS
(2)      COMPONENT frame_comp_gcd
(3)          GENERIC( width1 : integer);
(4)          PORT( clock, reset : IN std_logic);
(5)      END COMPONENT;
(6) ) END reuse_comp_library;
```

Table 10.2 Reuse component library containing frame components

Furthermore, this package or library can be used by any VHDL synthesis system. In our approach, a reuse component library is applied by our high-level synthesis system CADDY-II (Figure 10.4) ([Gutb91], [Gutb94]).

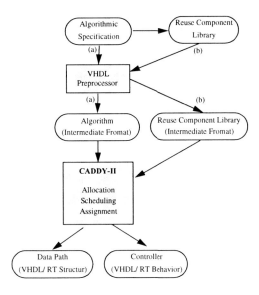

Figure 10.4 Synthesis flow using CADDY-II and the VHDL preprocessor

A VHDL preprocessor is used to execute the VHDL specific tasks. The preprocessor consists of one front-end, the VHDL parser, and two back-ends, one for synthesis specific tasks (*synthesis back-end*) and the second for simulation specific tasks (*simulation back-end*). The tasks of the simulation back-end are described in Section 10.6.

The synthesis back-end translates the VHDL specification into a specification described in an internal intermediate format (Figure 10.4 (a)). This format is used by the CADDY-II synthesis system representing a language independent input description. Furthermore, the synthesis back-end is used to transform the VHDL specification of the reuse component library into the same intermediate format

(Figure 10.4 (b)). This has to be done only if new frame components have been inserted in the library. Finally, CADDY-II can execute the usual high-level synthesis process. During this process, the operations of the algorithm are mapped onto components which provide the corresponding operators (Figure 10.1). Not only user defined operators, but also several standard operators can be used. The standard operators are defined in the IEEE *std_logic_arithm* package, and therefore, need not be specified in the reuse component library.

10.5 FRAME COMPONENT AND SYNTHESIS

For synthesis, the frame component contains a lot of important information. First, the relationship between procedures and the user defined component are determined simply by defining procedures and the corresponding component in the same frame component. Thus, for every procedure, the synthesis system is given information about the component to be mapped on. Next, synthesis specific information can be specified using VHDL attributes, e.g. number of gates, area, frequency or power. An example for specifying these attributes is given in Table 10.3. The same example of Table 10.1 is used extended by some required attributes.

Besides the attributes *Area, Power* etc. referring to the whole component, procedure specific attributes can also be specified. Using the attribute *Commutativity* it can be specified which input or output signals can be interchanged. The attribute *InitiationInterval* defines a period after which the component can be started again. Furthermore, this attribute is used to distinguish combinational and sequential implementations. If this attribute is set to the value '0', a combinational component is used, setting the attribute to a value greater than '0' causes a sequential component to be used.

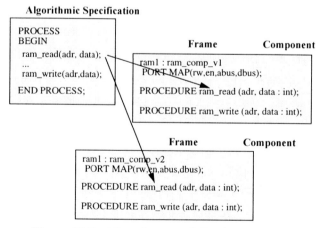

Figure 10.5 Allocating user defined components

The synthesis process can use the information concerning which procedures correspond to which component for an optimized allocation. Procedures that are contained in more than one frame component using the same name and implementing the same functionality offer the same optimization possibilities as standard procedures representing standard operators. In correspondence to the design constraints the synthesis system can now allocate the component with the best area/performance costs given by the attributes. A further possibility to determine the component to be allocated is that the designer explicitly specifies a reference to the corresponding component. This can be done by using the attribute *map_to_operator*.

```
(1)   ARCHITECTURE specification OF frame_comp_gcd IS
(2)      -- architecture declarative part
(3)      COMPONENT gcd_rt_comp
(4)         GENERIC(width1 : integer);
(5)         PORT( .... );
(6)      END COMPONENT;
(7)      ATTRIBUTE Gates OF gcd_rt_comp:
               COMPONENT IS 5 * width1 + 4;  -- XC4000 CLBs
(8)      ATTRIBUTE Area OF gcd_rt_comp:
               COMPONENT IS ((5 * width1 )/2+4, (5 * width1)/2);
(9)      ATTRIBUTE Power OF gcd_rt_comp: COMPONENT IS 1.5 * width1;
(10)     ATTRIBUTE Frequency OF gcd_rt_comp:
               COMPONENT IS 120 / width1;
(11)     SIGNAL start, done : std_logic;
(12)     SIGNAL x,y,z : integer;
(13)
(14)  BEGIN
(15)     gcd1 : gcd_rt_comp ...
(16)
(17)     PROCESS
(18)        -- process declarative part
(19)        PROCEDURE gcd(a : ...; b : ...; c : ...) IS
(20)           ATTRIBUTE Commutativity OF a : CONSTANT 1;
(21)           ATTRIBUTE Commutativity OF b : CONSTANT 1;
(22)           ATTRIBUTE Commutativity OF c : VARIABLE IS 2;
(23)           ATTRIBUTE InitiationInterval OF ALL : CONSTANT IS 0;
(24)        BEGIN
(25)           ...
(26)        END gcd;
(27)     BEGIN
(28)        -- stimuli
(29)     END PROCESS;
(30)  END specification;
```

Table 10.3 Frame component with synthesis attributes

In Figure 10.5, an example for operators mapping the read and write operations on several ram components is given. Using the procedures *read_ram* and *write_ram* in the specification, two different frame components exist containing these procedures. The synthesis system can now allocate the best fitting component by evaluating the attributes defined in both frame components.

10.6 FRAME COMPONENT AND SIMULATION

For simulation, the main aspects concerning the integration of user defined components in an algorithmic specification are: the communication between the access procedures and the RT components, and the implementation of simulation vectors. Note that emulation procedures does not cause any problems, because they can be used directly.

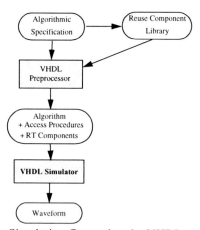

Figure 10.6 Simulation flow using the VHDL preprocessor

As mentioned before, the access procedures and components communicate via global signals with each other. This means that in every specification in which access procedures are used, the declarative part of the ARCHITECTURE has to contain the declaration of the user defined component as well as all necessary global signals. Furthermore, the implementation part of the ARCHITECTURE has to contain the component instantiation and the implementation of the procedure in the PROCESS declarative part. These hand-written descriptions are very time consuming. Due to the fact that the required information is already given in the corresponding frame component, the same preprocessor used for synthesis but the other back-end is applied to insert all necessary information in the specification. In this case, the back-end for simulation specific tasks (*simulation back-end*) is used. In contrast to the synthesis back-end, the simulation back-end generates a VHDL description instead of the language-independent intermediate format. The design flow containing our VHDL preprocessor is given in Figure 10.6.

A VHDL REUSE COMPONENT MODEL   141

```
(1)    LIBRARY ieee;
(2)    USE ieee.std_logic_1164.ALL; USE
(3)    ieee.std_logic_arith.ALL;
(4)
(5)    -- pragma DSL USE clib.reuse_comp_library.frame_comp_gcd;
(6)    ENTITY test IS
(7)       PORT(-- automatic extension of the interface
(8)             in1   : IN  signed(15 DOWNTO 0);
(9)             in2   : IN  signed(15 DOWNTO 0);
(10)            out1  : OUT signed(15 DOWNTO 0));
(11)   END test;
(12)
(13)   ARCHITECTURE example1 OF test IS
(14)      -- automatic insertion of global signals,
(15)      -- and component declarations
(16)   BEGIN
(17)      -- automatic insertion of the component instantiation
(18)      PROCESS
(19)         -- automatic insertion of procedure implementation
(20)         VARIABLE result : signed(15 DOWNTO 0);
(21)      BEGIN
(22)         gcd(in1, in2, result);
(23)         out1 <= result;
(24)      END PROCESS;
       END example1;
```

Table 10.4 Specification using an access procedure

Applying this preprocessor, the user has only to define the access procedures and a reference to the frame component to be used for simulation. During synthesis, this reference is ignored to allow optimization possibilities. In Table 10.4, a short example using an access procedure in a specification is given. The access procedure is called in line 21. The information to be inserted from the preprocessor are listed in the comments.

The component reference is specified using a pragma construct (Table 10.4 (line 4)). With this construct, the library, the package, and the frame component are defined in which the preprocessor can find the procedure and the corresponding component. For our preprocessor concept, several problems had to be solved:

- The user defined component may contain generic parameters, e.g. the width of one or several bitvector(s). In this case, the preprocessor extracts the necessary generic parameters from the given parameters of the procedure. Only, if this is impossible, the user has to define the missing generic parameters using a pragma clause.

- The preprocessor generates the needed global signals. In some cases, it is not possible to simply take over the name of the global signals defined in the frame component. Some of these names may already be used in the algorithmic specification. Thus, the preprocessor generates new names for those global signals.
- The correct reference, and the complete logical name of every global signal used in the procedure body are generated by the preprocessor. In the procedure body, several global signals can be used. To avoid possible name conflicts, these signals are used with their complete logical names.
- The preprocessor automatically extends the specification interface. This interface is extended by all present interface signals of the frame component (Section 10.3). The generic parameters of the frame component must not be inserted, because, in the specification, the actual values are used.
- The preprocessor automatically determines the relation between procedure and pragma construct. When several procedures are used in the specification which belong to different frame components, several relation conflicts may occur. E.g., two procedures *pa* and *pb* belong to two frame components *fa* and *fb* and are referenced by two pragma constructs. Now, procedure *pa* is contained in frame component *fa* as well as in frame component *fb*. First, the preprocessor uses the procedure parameter as indication for the correct relation. Assuming that the name of the procedure as well as the parameters are equal, the preprocessor suggests the most probable relation, but the user has to confirm or to correct this suggestion in order to generate a unique assignment for simulation.

The resulting VHDL preprocessor listing can be seen in Table 10.5. The reference *example1* is used to avoid possible name conflicts.

Until now the communication between access procedure and the corresponding component as well as the resulting preprocessor concept has been discussed. Next, simulation vectors have to be integrated. In our approach, in the frame component, simulation vectors for every procedure and the corresponding component (Table 10.1 (line 39)) can easily be inserted. The most appropriate place for these vectors is the implementation part of the PROCESS. Here, stimuli, which test the access procedure together with the component, can easily be inserted, as well as stimuli, which only test the component or an emulation procedure. As a result, "standard" simulation vectors are available at once, and they may be an important part in every further reuse and validation process [Hans97].

```vhdl
(1)   LIBRARY ieee;
(2)   USE ieee.std_logic_1164.ALL; USE ieee.std_logic_arith.ALL;
(3)   ENTITY test IS
(4)      PORT( -- automatic extension of the interface
(5)         clock, reset : IN std_logic;
(6)         in1, in2 : IN  signed(15 DOWNTO 0);
(7)         out1 : OUT signed(15 DOWNTO 0));
(8)   END test;
(9)   ARCHITECTURE example1 OF test IS
(10)     -- automatic insertion of the component declaration,
(11)     COMPONENT gcd_rt_comp
(12)        GENERIC(width1 : integer);
(13)        PORT(clock, reset : IN std_logic;
(14)           start : IN std_logic;
(15)           done : OUT std_logic;
(16)           x,y : IN signed(width1-1 DOWNTO 0);
(17)           z : OUT signed(width1-1 DOWNTO 0));
(18)     END COMPONENT;
(19)     -- automatic insertion of the global signals
(20)     SIGNAL start, done : std_logic;
(21)     SIGNAL x,y,z : signed(15 DOWNTO 0);
(22)  BEGIN
(23)     -- automatic insertion of the component instantiation
(24)     gcd1 : gcd_rt_comp
(25)        GENERIC MAP(16);
(26)        PORT MAP (clock, reset, start, done, x,y,z);
(27)     PROCESS
(28)        -- automatic insertion of the procedure declaration
(29)        -- and implementation
(30)        PROCEDURE gcd(
(31)           a : IN signed(width1-1 DOWNTO 0);
(32)           b : IN signed(width1-1 DOWNTO 0);
(33)           c : IN signed(width1-1 DOWNTO 0)) IS
(34)        BEGIN
(35)           example1.start <= '1';
(36)           example1.x <= a;
(37)           example1.y <= b;
(38)           WAIT UNTIL example1.done'event
(39)                  AND example1.done = '1';
(40)           example1.start <= '0';
(41)           c := example1.z;
(42)           WAIT FOR 1 ns;
(43)        END gcd;
(44)        VARIABLE result : signed(15 DOWNTO 0);
(45)     BEGIN
(46)        gcd(in1, in2, result);
(47)        out1 <= result;
(48)     END PROCESS;
(49)  END example1;
```

Table 10.5 VHDL specification after preprocessing

10.7 CONCLUSION

This contribution presented a new approach for designing VHDL models for mixed abstraction level simulation and behavioral synthesis. The advantages of the presented approach are, first, that the concept of the frame component offers all necessary properties for mixed abstraction level simulation and synthesis. Concerning simulation, the access procedures or emulation procedures, and the corresponding user defined component are contained in a frame component. Thus, our VHDL preprocessor can optimally support the designer applying access procedures in an algorithmic specification, and executing the final simulation. Concerning synthesis, specific attributes can be defined to extend the further optimization potential during the high-level synthesis process. Second, the user defined components can be implemented at any arbitrary abstraction level supported by VHDL. Third, a reuse component library can easily be implemented using the VHDL library concept. Finally, our approach is based on the IEEE VHDL standard and can therefore easily be used by any VHDL simulation or synthesis tools.

11
VIRTUAL COMPONENT INTERFACES

M. M. Kamal Hashmi

Design Automation Centre
ICL, Wenlock Way
Manchester, U.K

11.1 INTRODUCTION

When selecting a Virtual Component (VC), the user initially depends on a high-level specification of the VC to match with his functional requirements. However, for final selection and to actually integrate the VC, the user needs a specification of its interface. In order to maximize the chances of choosing correctly, this interface must be unambiguous and complete, specifying not only the static interface to the VC, but also the dynamic interface i.e. its communications protocol.

This contribution describes why such interfaces should be an integral part of a System-On-Chip methodology and how they can be used in a design flow. In Section 11.3 and Section 11.4, a methodology is described where the rigorous separation of interface from function, by providing an explicit interface specification, considerably eases the re-use of VCs. To aid the VC user's comprehension and to enable property-consistent modelling at different levels of abstraction, the protocol should be clearly defined at these multiple levels – and the greatest benefit to the user is given by the executable specification of such information.

In Section 11.5, the rationale for the interface-based design methodology and its advantages for the concurrent engineering of large projects, for design decomposition and for verification strategies are also described.

Finally, in Section 11.6, Section 11.7 and Section 11.8, the contribution relates these concepts to the current work of the VSIA SLD Interface subgroup and DASC Systems and Interface Design Working Group and their standardization efforts.

11.2 DEFINITIONS AND TERMINOLOGY

Intellectual Property (IP) components for which there may, or may not, be a physical realization are termed Virtual Components (VCs). These can be as general as algorithms or constraint equations, and as specific as an RTL model or net list of gates.

11.3 VC CREATION AND REUSE

Virtual Components are initially called into existence through their specifications, which may be in a natural or a formal language and may be executable as a model of the VC. All VCs must have an implementation, at some level of abstraction, in some form – soft, firm or hard – if they are to be used in a completed system. So already we see that a VC must have at least two views or levels: a specification and an implementation.

A user composes a system by choosing VCs from a set of specifications and/or specifying a set of VCs, which in their turn may use other VCs. Of course, VCs specified in a natural language are less easy to use than those specified unambiguously in a formal language. VCs with executable models, as well as a quickly comprehensible specification, are the cheapest to re-use. So an important factor in maximizing re-use and minimizing design effort is the cost of creating VCs with "good" specifications – formally captured and verified, executable and "trustworthy". To analyze this cost we need to look at VC sources. There are four basic sources for VCs:

1. The user may design a VC.
2. A VC may be created by another user or team.
3. A VC may be re-used from another or earlier projects.
4. A VC may be bought in from an external source.

1. VCs designed by the user are the most trusted by the user – and are usually the worst kind of VC for re-use by another user! These VCs often do not even have an accurate specification since the user can change or ignore his own specification because he is designing both the *used* VC and the *using* VC. It needs very disciplined design principles in such a case to maintain the specified functional partitioning between the components.

2. VCs designed by another part of a project are also trusted for integration by the user, since they are often created to the user's specification. However, problems can arise when the specification is incorrect or ambiguous or incomplete. In this case, good communication between teams is essential for the project to be successful – and when this communication is not via the VC specification then this specification is less accurate and its re-use capability suffers.

3. VCs already created by other projects rarely have accurate specifications. Since a project usually has to minimize its costs in time, money and resource, its main focus (especially towards the latter stages) is on the implementation and not on the specifications, so VC specifications are not kept up-to-date or completed – just enough is done to allow implementation to continue and fill the gaps. Therefore there is an extra cost in making the VCs re-usable by other projects – and this is usually done as a post-project process and hence is often omitted.

There are efforts in companies, for example Alcatel [Delf98], to create internal economies which offset such costs to the project, but these are rare and usually add an overhead to the overall cost which is hard to justify to managers.

4. Most VCs successfully bought in from external sources are usually simple and cheap. Complex VCs are rarely bought in – not just because of the technical problems of complete and accurate specification, integration and verification but mainly because of the legal and commercial difficulties [Tamm98] of externally sourcing black box components.

Generally, external VCs are only used where the integration overhead is significantly less than the design cost of the VC i.e. the cost incurred due to the complexity of the interface is swamped by the savings from not implementing the complexity of the functionality. Such VCs are probably more accurately categorized as 'subsystems' – examples include processor and DSP cores. Even so, successful examples of business with these kind of cores are rare, and are characterized by relatively good documentation and support services.

11.4 CUTTING THE COST OF REUSE

The main technical barrier to obtaining easily re-usable VCs in the first three cases described above is the accurate maintenance of the specification through the evolution of the design. In the first case, the specification is not essential for implementation; in the second, it may contain errors and is again not essential for implementation; and in the third case, the specifications have to be updated after the project has been completed in order to be able to be re-used.

A design methodology in which the specifications of the VCs are actively used during the design and implementation phases and hence maintained throughout the design cycle would considerably cut the post-project cost of VC re-use. If this methodology also helped the design and verification of complex systems then it would also be seen by the engineers during the project as an asset – rather than an overhead.

Methodologies rooted in Interface-based Design ([Hash97] [Rows97]) and using multiple levels of abstraction ([Sang96] [Hodg97]) have been shown to considerably improve both the design and the verification processes ([Wilk99] [Hash95]).

One of the key components of Interface-based Design is the separation of the interface from the functionality. The Interface Specification contains the specification of both the static and the dynamic interface to the VC – it should specify not just the ports and data types on the VC interface, but also the communication protocol used by the VC through the ports, including implicit dependencies between ports.

It becomes obvious that such an interface specification provides a very useful black-box specification of a VC. And since it is part of the executable specification of the VC and hence part of the verification model, it is used and maintained during the project life cycle and remains an accurate specification of the VC.

So the use of an interface-based design methodology during the design, implementation and verification of a project goes some way to solving many of the technical problems of VC re-use. In the first case above, using a methodology that separates the interface specification from the functional implementation means the designer has to maintain the partitioning between VCs – or update the interface specification if the functional partition is changed. In the second case, exchanging and sharing interface specifications between teams guarantees, early in the design cycle, that the agreed interface is unambiguous and verifiable while allowing the teams to work separately at their own pace [Hash95]. In the third case, after a project is completed the interface specifications are guaranteed to be accurate and verified, enabling easier VC selection. Also in this case – even with accurate specifications – VCs often are not reused because, although the function may be correct, the interface protocol is not the required one. However, VCs designed using an interface-based methodology are much easier to customize to a different protocol since the communication mechanisms are easier to separate. Interface-based design even helps in the fourth case - interface specifications can be more freely exchanged for VC selection and evaluation without the worry of 'IP pollution' since the functional implementation is separate and the interface specification does not give away any of the implementation.

11.5 INTERFACE BASED DESIGN

What are the other reasons mentioned above – which aid the design and verification process itself – for moving to an interface based design methodology?

One of the commonest ways of managing complexity is by using *divide-and-conquer* methods. And the simplest way of implementing such a methodology in a design is the partitioning of a design into components which, in their turn, may be further partitioned – leading to the traditional hierarchic model of design units.

However, this partitioning won't make life any easier unless – at each level of the hierarchy – each component hides all its internals and only makes visible what is needed. In this way, the design task at each hierarchic level is minimized – only the component itself needs to worry about its internal implementation.

The specification of the visible part of the component – all the parts of its design which affect how it is used – is usually called its interface. However (as we shall

VIRTUAL COMPONENT INTERFACES

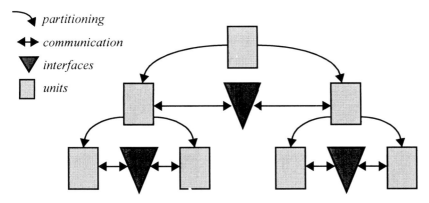

Figure 11.1 Hierarchic partitioning with interfaces

see) interfaces can also be defined between components, to specify their interdependence. Both the unit and inter-unit interfaces are needed to define the full 'black-box' specification of a component. Units should not communicate directly with each other – all communication is via the interfaces used by the unit (see Figure 11.1).

11.5.1 Static Interfaces plus Communication Specification

Traditional interfaces in the HDL world are generally designed to allow the structural composition of components without typing or size errors. So the basic attributes defined would be the number of ports, their names (or the ordering), the sizes in bits of the ports and the port direction. Additionally, some component interface definitions specify a port type which can be checked.

This kind of interface enables the user to build a structure or network of components without gross errors. However he will not know if the structure makes any kind of sense – or even if two components that his network connects without any apparent errors can successfully exchange data. Some extra information is required for data exchange to be verifiable in compositions and for a true black box description.

One of the crucial extra bits of information needed is how the component ports are used to exchange information. This data is currently usually provided as comment text and timing diagrams describing the components communications protocols (see Figure 11.2).

So you can see that the usual HDL interface is not a true specification at all - we need not just the port types and directions but also how these ports are used to send and receive data. Extra information is needed to describe the communications protocol – the syntax of the conversations in which the VC is capable of participating.

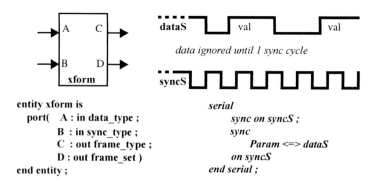

Figure 11.2 Static interface + dynamic interface

11.5.2 Protocol Definitions between Units

When a component talks to more than one component then it usually has separate protocols governing communications from different sets of ports to the other components. So the overall unit interface for the component should ideally provide the full list of protocols that it uses to all possible other components. Each of these protocols would define a conversation syntax.

Why would we want to have such a protocol definition? Isn't this just design repetition since at some point we will have to implement the protocol anyway? The answer is that there are three areas in which the separate definition of communication proves greatly useful:

1. *Design* – designing functionality separately from the communication mechanisms enables the design to progress at a faster rate because of the separations of concerns: The design of the required communication is separated from the design of the required functionality and from the implementation of the communication. Also, this separation of concerns means that the high level design phase of 'what-if' modelling is much less onerous, much clearer and re-iterates to a satisfactory conclusion quicker.

2. *Specification* – having an unambiguous specification of the total communication between any set of functional units allows you to partition the design between teams more easily. The teams can independently develop their parts of the overall design secure in the knowledge that when they connect the models together, there will be no communication errors. This is a very large management benefit, since inter-team communication in large projects is always a big problem. The common interface specification can only be changed by agreement and any changes and their dynamic and static effects are clearly visible.

3. *Reuse/Componentisation* – the separation of the functional core from the communication mechanism allows much greater design reuse - not only can a different communication mechanism be slotted in place, the communication protocol itself can be reused elsewhere. Additionally, the pure functional core is a more easily reusable component than when it is combined with its communication mechanism. Naturally, given the nature of the interface protocols, component selection is improved – this applies to the design levels as well as the implementation levels.

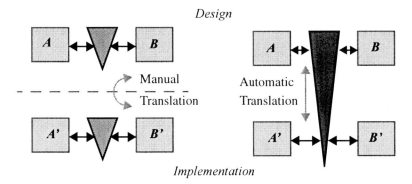

Figure 11.3 Working at multiple levels of abstraction

11.5.3 Multi-level Interfaces

So far, we have only considered interfaces as pure communication protocol containers. However, an inescapable factor in producing large complex systems is that to design such a system in a reasonable time with no errors, multiple levels of design must be used. At each level, a different design problem is addressed and this separation of concerns allows the designer to design a more complex system in a shorter time than he would otherwise have been able.

In traditional HDLs, the use of multiple levels of design in a system leads to different interfaces between units at different levels. This in turn leads to separate execution models for each level. The models at one level cannot usually communicate with the models at another level because the interface protocols are different.

However, if the interface container had not just the protocol at one level specified within it, but the protocols at all levels for which models existed then the interface could automatically translate between the levels (see Figure 11.3).

The ability to build and use mixed multilevel models (**mMm**) has some extremely useful consequences:

- Full System Models can be built much earlier – as soon as models at any level (instead of all at the same level) are available. Additionally, the more detailed models at lower levels can be used and simulated in a system model as soon as they are available.
- Testbenches written for units at a higher level can be reused to test the lower level units as well. The only extra testing that needs to be done is that which verify the additional detail which is added at the lower level.
- Much faster simulation – this sort of mixed-level simulation, where one of the models is at a low level and the others are at a higher level, tends to be much faster since higher level models have less detail and thus run faster.
- Token-based performance models can be co-simulated with real data models since the level translation can discard data going up to token-based level and generate it going the other way (either from random data, or by calling a user supplied function).

All of these capabilities of multilevel interfaces mean that more verification can be preformed at each and every level. And since errors found at higher levels are much easier to fix than if they were found later in the design cycle, this enables the whole design and verification cycle to proceed much more rapidly.

11.5.4 Using Interfaces with Functional and Architectural Models

A major source of confusion in how to use interfaces is usually caused by the dichotomy and gap between the models (algorithmic or otherwise) at the functional level that are used to model and design the overall functionality, and the models at the architectural level that are used to investigate and choose implementations based on different architectural patterns.

While writing the functional model, interfaces speed up the design process by allowing the separation of function and communication. Moreover, interface protocols from previous communication designs can be reused or enforced (if part of a design standard). So, in the functional model, you will eventually end up with a detailed partitioned functional design and the specification of the communication protocols between the different parts of the model.

These protocols are then useful when designing the architectural model since they have also to be implemented, and the information and control transfer provided by the protocols are invariant between the two different kinds of models.

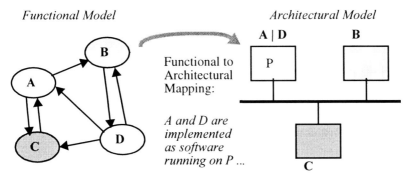

Figure 11.4 Bridging the Gap between the Function and the Architecture

The architectural model is a mapping of the functional model into a chosen architecture. Initially the mapping has little additional implementation detail being added because the feasibility of the architectural pattern with the designed functionality is first being verified – the size fit and the performance needs to be checked. When the performance is satisfactory – either by adjusting the architecture or the functional design – then the additional required implementation detail is added to the model. The continued validity of the models during this process is made easier to verify by being able to map the interfaces.

Interfaces at the architectural level can use newly designed protocols at the wire or bus level but most commonly use standard protocols. The mapping of the functional parts of the model to architectural components requires that the interfaces between the functional parts must be able to be mapped to the interfaces between the architectural components. In other words, the chosen architectural level interface protocols must be able to handle the designed communication between the components.

So it becomes obvious that one of the first important tasks in choosing an architectural pattern is to add an architectural level to the – up to now – pure functional interfaces. Depending on the architecture, the functional interface could be mapped to a hardware-hardware interface, or to a hardware-software interface, or to a software-software interface. It should be emphasized that this mapping **must** be possible if the designed functionality is to be implemented - and so the mapping is an initial check on the feasibility of the design.

And, once the mapping has been done, the performance and design correctness of the architecture can be checked throughout the implementation phase because the multi-level interfaces allow you to mix architectural and functional components in a simulation.

11.6 VCI SPECIFICATION STANDARDS EFFORTS

So, from the introduction above, it is apparent that the key to design-for-re-use and VC exchange is interface-based design; and the key to interface-based design is the Interface Specification or Virtual Component Interface (VCI).

11.6.1 The VSIA SLD Interfaces Subgroup

The Virtual Socket Interface Alliance (see http://www.vsi.org) is chartered to define, develop, authorize, test and promote open standard specifications relating to data formats, test methodologies and interfaces.

The VSI System Level Design development working group recognized the importance of full interface specifications and set up an Interfaces subgroup to look at interfaces so that VCs could easily be exchanged and used. This group is not concerned with defining a standard interface for VCs - unlike the VSI On-Chip Bus (OCB) group - but with defining standards for specifying interfaces.

11.6.2 The IEEE DASC SID Working Group

Ideally, interface specifications should be in a formal language which can be verified dynamically or statically, and should specify the interface at all the levels of abstraction used by the VC provider and needed by the VC user. Automatic translation between the levels would enable the VC user to use the VC in verification models at different levels of abstraction, while allowing the VC provider to write models at only the essential levels.

A language is being developed to perform this task and its current incarnation is as an extension to VHDL [IEEE93], called VHDL+ [Hash00]. The extensions are being considered for a standard by the DASC Systems and Interface-based Design (SID) working group (see http://www.eda.org/sid).

In principle, the interface extensions in VHDL+ are not language specific and could be used, with the same or different syntax, for other languages like Verilog, SDL or C/C++; or for specifying interfaces between models in these languages.

11.7 VSI SLD INTERFACE DOCUMENTATION STANDARD

Currently, documentation provided for VCs mix up the interface and the function such that it is difficult to quickly understand exactly what the VC does, and also very difficult to work out how to communicate with the VC if it is chosen. Usually, some waveform diagrams are given for a few specific transactions. If it is an advanced specification, perhaps some Message Sequence Charts (MSCs) will be given. Whereas detailed timing characteristics for each port will be specified in an appendix, little help will be given to understand the protocol.

The VSI SLD Interfaces Subgroup is writing a System-Level Interface Behavioral Documentation Standard (SLIF) [VSIA00c] which will enable interfaces to be documented in a standard way. The documentation standard also helps to ensure the completeness, uniformity and clear hierarchical description of VC interfaces at different levels of abstraction.

To explain how the standard specifies the static and dynamic aspects of an interface at different levels of abstraction, we need to start with how it looks at interfaces (see Figure 11.5).

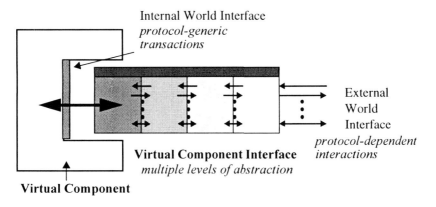

Figure 11.5 The Virtual Component Interface

The VCI is a component that mediates between the VC and the external world. If the abstraction levels of the internal and external worlds are not the same then the VCI provides the specification of the mapping between the two levels. This mapping may go through intermediate levels of abstraction.

Components communicate through ports and via channels to other components by using a selection from a set of objects ranging from transactions and messages to packets and cells. All these objects can have attributes that specialize the object – for example a transaction can be buffered or non-buffered, blocking or non-blocking. The general object types are documented in an Interface Taxonomy document being co-written by the Interfaces subgroup and the OCB group, and the attributes are documented in the SLIF. For each layer of abstraction, the static and dynamic mappings from it to the next higher layer is also defined - the highest level of communication abstraction being the no-time, infinite-resource pure functional level.

The rough structure of the interface layer portion of the document for each level of abstraction is as follows:

156 VIRTUAL COMPONENTS DESIGN AND REUSE

- Layer Identification, and Interface Name:
  - Structural Description
    o Port Identification
    o Inter-Layer Static Mapping
  - Behavioral Description
    o Attributes
    o Transactions
    o Inter-Layer Behavioral Mappings
    o Protocol Description

So, for each level of abstraction, there is a separate section. This section starts with a description of the level, followed by sections on the structural and behavioral descriptions. The Structure subsection defines the ports and the types (which are defined in an earlier layer-independent section) and then defines the inter-layer static type mappings if needed. The Behavior subsection starts with a specification of the port attributes and the transactions through the ports, followed by the specification of the inter-layer behavioral mappings if needed. Finally there is a description of the protocols used by the ports and transactions. All these descriptions are, as far as possible, independent of any formal language.

The Interface Behavioral Documentation Standard was released by the VSIA in March 2000 and is available from the VSIA web site (*www.vsi.org*). It is being validated by pilot project use now.

11.8 VHDL EXTENSIONS FOR INTERFACES AND SYSTEM DESIGN

Interface-based Design was first tried out in ICL in the early 90's using an internal language called CHISLE [Jebs93]. This language captured clocked interfaces between units at multiple levels of abstraction and performed automatic translation between the levels. To allow more designers to use the methodology it was decided to re-implement it based on a standard HDL and to further extend it to be able to capture more general types of interfaces.

VHDL+ is a pure superset of VHDL, adding new features to the language without removing any of the current capabilities. The main additions to the language include a new primary object called an Interface, a new tertiary object for architectures called an activity and the ability in architectures to use Interfaces by using send and receive statements.

Interfaces define a communication protocol that is used by one or more ends. The protocol tree and patterns are defined in the Interfaces objects. Entities connect with interfaces by using a specific end of the interface channel. Interface specifications contain a collection of transactions, messages and signals that can be composed into higher levels of abstraction using serial and parallel blocks and temporal abstractions. The composition is a declarative construct and so allows the translation between levels of abstraction in either direction - up or down the levels (see Figure 11.6).

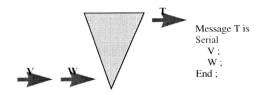

Figure 11.6 Interface level translation – example

Interfaces in VHDL+ provide a complete static and dynamic specification for the communication protocol between units. They also enable the effective use of multi-level modelling across teams working at different levels, and considerably increase the amount of verification possible by allowing testbenches written at a high level of abstraction to test low-level units.

11.9 CONCLUSION

Interface-based Design is a methodology that considerably eases the task of design for re-use. It also enables a much better way of specifying VCs by allowing interfaces to be captured separately from functionality at different levels of abstraction during design.

The VSIA Interface documentation standard combines such Interface Specification with necessary documentation and the specification of other facets of the VC. It will provide a standard way of exchanging VCI information quickly and comprehensibly.

So, an interface written in VHDL+, combined with some graphics (like MSCs) and documentation like the SLIF, would make a very effective (executable and comprehensible) multi-level specification for a VC. The main deficiency currently is the lack of constraints specification and this can be remedied by combining its use with a constraints language.

11.10 ACKNOWLEDGEMENTS

The author would like to thank the VSIA, all the contributors to the VSIA SLD DWG and especially the members of the SLD Interfaces subgroup who include (but are not limited to): Chris Lennard, Brian Bailey (now chair of the subgroup) and Gjalt de Jong.

12 A METHOD FOR INTERFACE CUSTOMIZATION OF SOFT IP CORES

Robert Siegmund and Dietmar Müller

Chemnitz University of Technology
Professorship Circuit and System Design
Chemnitz, Germany

Abstract

In this chapter, we present an interface customization technique suitable for Soft IP cores, which are to be implemented as synthesizable VHDL models. By exploiting the features of VHDL+, an extension to VHDL, we separate the specifications of the IP functional behavior and IP interface protocols into two distinct IP design units. Interface customization is done through specification of system specific IP interface communication protocols, followed by an IP interface generation at signal level with our tool MODIS.

12.1 INTRODUCTION

In microelectronic systems design, block oriented reuse is being considered a necessity today in order to successfully implement million-gates SoC under the pressure of short time-to-market constraints. Therefore, techniques need to be developed that enable an efficient adaptation of existing hardware blocks, referred to as IP or *Intellectual Property* blocks, to new specifications. Figure 12.1 visualizes the IP selection and instantiation flow in IP based system design. From the system specification, a set of requirements R on IP function and interface and a set of IP constraints C are derived. These requirement and constraint sets will be formally captured in an IP specification which is then used to (automatically) retrieve a corresponding IP implementation from a repository. It will be often the case that there is no IP in the repository that matches the specification exactly, so that an

160   VIRTUAL COMPONENTS DESIGN AND REUSE

adaptation of function and/or interface of the retrieved IP to the specification is required before it can be integrated into the target system.

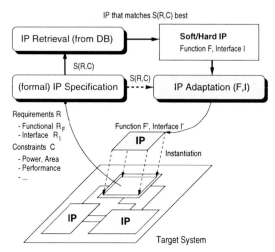

Figure 12.1 IP selection and instantiation process in IP based system design

In this chapter, we want to focus on the problem of interface customization of Soft IP cores. Hardware Soft IP are usually implemented as technology independent, synthesizable HDL models. Common HDL's, such as VHDL or Verilog HDL, require the specification of the IP interfaces at physical signal level with a fixed signalling protocol. This is a potential obstacle for further reuse since different target systems may impose different interface requirements on the IP such as the data communication protocol to be used or the physical implementation of the communication paths (e.g. uni/bidirectional buses).

Interface customization of an IP module that is provided as an HDL model is currently done through either manual modification of the HDL source or through design or generation of 'glue logic' between IP and system environment in form of a protocol conversion module ([Pass98], [Smit98]).

Manual modification of the HDL source by the IP user is the most straightforward approach but requires detailed knowledge about the implementation of the module, which contradicts IP protection demands. Moreover, in HDL descriptions, module behavior and module communication are more or less intertwined, making manual customizations of the module interface a tedious and error-prone task. Alternatively, a set of different interface implementations may be provided with the source code from which one is selected by means of a HDL preprocessor. The disadvantage of this method clearly is that the module interface can only be customized for a limited number of communication schemes.

[Pass98], [Smit98], [Nara95], [Bori88], [Akel91] describe algorithms for automatic generation of protocol conversion modules in form of FSM's that translate

A METHOD FOR INTERFACE CUSTOMIZATION OF SOFT IP CORES   161

between incompatible protocols. The use of a protocol conversion module has the advantage that a modification of the module implementation is not necessary. However, additional hardware is required to implement the conversion module. As a consequence, data transfer latencies between system modules as well as chip area and power consumption of the system implementation may increase considerably.

In this chapter, we propose a design for reuse methodology focusing on customizable synchronous interfaces of IP soft cores that are implemented as synthesizable VHDL specifications. Our approach is to follow the ideas presented in [Rows97] and to separate the specifications of IP core functional behavior and core interface, that describes the protocol used to communicate with the IP, into different design units. The interface unit can then be easily adapted to system specific communication protocols without the need to modify the module behavioral description. Furthermore, we developed a tool called MODIS, which can be used to synthesize an implementation of the IP core interface from the interface protocol specification.

12.2 METHOD

The method described in this chapter is applicable for the implementation of Soft IP cores with synthesizable VHDL+ models at RT or algorithmic level.

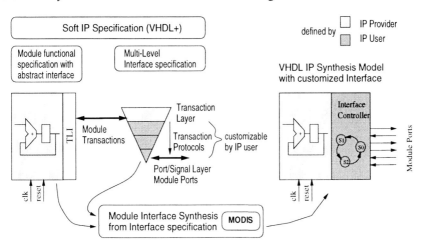

Figure 12.2 Principle of IP interface customization

At the system level, a hardware module can be thought of as an object that communicates with other objects of a system by means of abstract transactions. Transactions describe the movement of information between system modules, such as data, addresses or control sequences. For the specification of the module functional behavior, it is of interest to know which transactions the module should perform in order to communicate with the system environment. However, the mechanism of data transfers during transactions, e.g. the protocol to be used and the physical implementation of the communication channel, are specified by the system in which the module is to be integrated and therefore should be abstracted from the IP behavioral specification. Figure 12.2 visualizes our approach to specification and synthesis of customizable IP interfaces. The IP model is divided into a specification of the module functional behavior which uses a Transaction Level Interface (TLI) to communicate with its environment via high level messaging, and an interface specification that defines the set of transactions that can be performed by the IP. The interface unit furthermore defines a hierarchically structured protocol for each transaction and a set of module ports that form the physical implementation of the module interface. It is left to the IP user to customize the transaction protocols and module ports of the interface specification for a particular communication scheme.

As formal modeling language for Soft IP cores we use VHDL+ [ICL98], which extends VHDL with concepts of high level messaging and interface-base design [Sieg98]. In order to generate synthesizable VHDL code from the VHDL+ IP model the tool MODIS was implemented. MODIS generates from the interface specification an implementation of the IP interface in form of an interface controller, that replaces the TLI of the functional specification of the IP. The output of MODIS is a synthesizable VHDL model of the Soft IP core with a customized, system-specific interface implementation.

12.2.1 Interface Specification

The interface unit of the IP model declares the transactions that the IP can perform in order to communicate with its environment. Each transaction is specified by a unique identifier and a set of parameters forming the 'payload' or the information content of the transaction. Each parameter has an additional attribute that indicates the direction of the information flow. The example code in Table 12.1 shows the interface specification of an IP (for demonstration purposes we chose a hypothetical CPU core as example) at transaction level in VHDL+ syntax. The transactions **io_write** and **io_read** transport a data byte from the CPU to a system address and vice versa. At this stage the module interface is completely specified at transaction level. This interface specification forms in combination with the later described functional specification of the CPU an executable VHDL+ model of the IP that may be simulated and verified with the SuperVISE simulator [ICL98] and checked into an IP repository.

In order to customize the IP core interface for a system-specific communication scheme, the interface specification has to be extended and refined by the IP user. For each transaction in the interface a *composition* is to be defined that describes the transaction protocol. A transaction composition consists of a sequence of VHDL+ interface statements such as instances of other transactions or messages[1],

---

1. Messages are a special case of transactions where all parameters have the same direction property

A METHOD FOR INTERFACE CUSTOMIZATION OF SOFT IP CORES   163

which must be declared in the interface and which in turn contain a composition. This enables the specification of hierarchically structured IP interface protocols with an arbitrary complexity. At the lowest protocol hierarchy level a mapping of transaction parameters to interface signal values using VHDL+ signal associations (denoted with the '<=>' symbol) is defined. Interface signals describe the physical implementation of the communication channel and specify the module ports that are inferred during interface synthesis (see Section 12.3).

```
interface cpu_if is

   between cpu, system;
   protocol is
      io_write; io_read;
   end;

   transaction io_write(
      addr: byte from cpu to system;
      data: byte from cpu to system)
   is between cpu, system; end;

   transaction io_read(
      addr: byte from cpu to system;
      data: byte from system to cpu)
   is between cpu, system; end;

end interface;
```

Table 12.1 IP interface specification at transaction level

For specification of synchronous interfaces, the mapping of transaction parameters to interface signals is synchronized to an interface clock signal using the VHDL+ **sync**-statement. Table 12.2 lists the protocol specification of transaction 'io_write', for which a 2-wire serial protocol has been implemented. In this example, the transaction uses a physical channel that consists of two unidirectional lines SDO (serial data out) and SOF (start of output frame). The transaction parameters 'addr','data' are transmitted sequentially bit for bit, starting with the MSB of parameter 'addr'. Start of transmission is indicated by a '1' on the SOF-line. An arbitrary number of idle cycles may precede the transfer. An important feature of the interface specification is that it is purely declarative, e.g. it specifies the interface signal sequences that are valid for this particular protocol rather than a corresponding behavior that produces or consumes these sequences. Therefore, the level of abstraction for the interface specification is effectively raised and the modelling effort is reduced.

```
transaction io_write(addr: byte from cpu TO system;
                    data: byte from cpu TO system)
  is between cpu, system;
  composition
     variable idle_cycles: integer;
  serial
     for i in 0 to idle_cycles loop
        map_physical_out('-','0');
     end loop with map_physical_out(addr(7),'1') from cpu;

     for i in 6 downto 0 loop
        map_physical_out(addr(i),'0') from cpu;
     end loop;

     for i in 7 downto 0 loop
        map_physical_out(data(i),'0') from cpu;
     end loop;
end;

message map_physical_out(s_do, s_sof : bit)
   is from cpu to system;
   composition serial
   sync
      SDO <=> s_do;
      SOF <=> s_sof;
   when rising_edge(BUSCLK);
end message;

-- Declaration of physical communication
-- channel (2 wire serial bus)

signal BUSCLK,SDO,SOF : bit is from cpu to system; end;
```

Table 12.2 Specification of a 2-wire serial protocol for transaction

12.2.2 Specification of the Module Behavior using Interface Transactions

The module functional behavior is specified in form of a synthesizable[1] native VHDL model. In order to perform high-level communication via the TLI using the transactions that are defined in the interface unit, we extend the VHDL synthesis

---

1. with respect to the SYNOPSYS VHDL synthesis subset

A METHOD FOR INTERFACE CUSTOMIZATION OF SOFT IP CORES   165

subset with the VHDL+ *transaction-join* statement[1], which enables a VHDL behavioral model to initiate or participate in an interface transaction.

The VHDL+ **transaction join** statement may be used in VHDL processes like any VHDL sequential statement with the restriction that the enclosing processes must not have a sensitivity list. This is due to the fact that, like the wait-statement, these statements do not execute in zero time but will block the process execution until the transaction has completed. Concurrent use of a particular transaction (e.g through instantiation with transaction join statements in concurrent processes) is not allowed, because transactions represent a resource (rather than a procedure) that is limited to one instance.

```
-- CPU Entity with TLI
use interface work.cpu_if;

entity CPU is
  port(clk  : in bit;
       reset: in bit);
    interface port(t_if: cpu of cpu_if);
end CPU;

-- CPU Architecture
architecture VHDLPlus_TLI of CPU is
begin
    instruction_fetch: process
    begin
        -- fetch instruction
        join t_if.io_read(pc, instr);
        pc := pc + "1";
        ...
    end process instruction_fetch;
end VHDLPlus_TLI;
```

Table 12.3 Behavioural specification using a TLI and interface transactions

The code example in Table 12.3 is extracted from the behavioral specification of the CPU. The entity declaration of the module is extended with the VHDL+ **interface port** clause that represents the TLI of the module. Furthermore does the top level entity declaration contain all clock and reset signals that are used in the behavioral model of the IP. With the VHDL+ **use interface** clause the interface specification to be used is selected. From the CPU behavioral specification, a fragment of the instruction fetch process is shown, where the 'io_read' transaction is used to fetch an instruction from the address given by the contents of the variable 'pc' and place it into variable 'instr'.

---

1. during interface synthesis, this statement is replaced with VHDL code

12.2.3 Interface Customization of conventional Soft and Hard IP

Soft IP that are implemented as VHDL or Verilog synthesis models as well as Hard IP do not have a TLI. In order to integrate these IP in our approach, we suggest to use a VHDL+ interface wrapper that abstracts the IP interface at transaction level

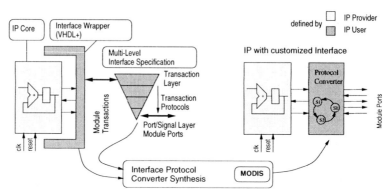

Figure 12.3 Interface customization for Soft- and Hard IP without TLI

(Figure 12.3). Now the interface specification and synthesis method as described in Section 12.2 can be applied. In this case, the output of the interface synthesis is a protocol converter that translates between the data communication protocols of IP interface and system environment.

12.3 INTERFACE SYNTHESIS

Goal of the interface synthesis is (1) to generate from the interface specification an implementation of the IP core interface in form of an interface controller, and (2) to replace the TLI of the IP core model with this controller (see Figure 12.2). The synthesis output is a synthesizable VHDL model of the IP core with a customized, target system specific interface implementation.

12.3.1 Generation of Interface Controller Machines

Inspired by the work in [Pass98], [Ober96] we express the protocol of each transaction that is specified in the interface by means of a regular expression r_p. R_p describes the set L_p of legal signal sequences at the module ports for a transaction using this protocol. A symbol σ in the alphabet Σ of the regular expression r_p is specified as the n-tuple of signal values that a subset of the module ports take on

A METHOD FOR INTERFACE CUSTOMIZATION OF SOFT IP CORES 167

at time t^k of the corresponding interface clock. The protocol of transaction **io_write** in Figure 12.2 could be expressed with a regular expression as follows:

$$\sigma_k = \{SDO(t_{BusCLK}), SOF(t_{BusCLK})\}$$

```
r(io_write) =
   <  '-'    ,'0'>*  .
   <addr(7),'1'>  .  ...  .  <addr(0),'0'>  .
   <data(7),'0'>  .  ...  .  <data(0),'0'>
```

We developed an algorithm that constructs a regular expression for each TLI transaction according to Table 12.4, representing the transaction protocol. Instances of other transactions and messages in a transaction composition are resolved so that eventually the composition is a sequence of VHDL+ **sync** statements, representing symbols σ_k, and regular expression operators '.' (concatenation) and '*' (zero-more-closure). Our algorithm then builds a non-deterministic finite automaton M

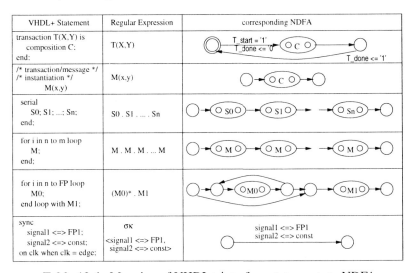

Table 12.4 Mapping of VHDL+ interface statements to NDFA

for each regular expression. Next, to M an equivalent deterministic finite automaton M' is determined using the subset construction algorithm [Hopc86]. At this point, all edges E of the state transition graph STG_M of M' are labelled with signal associations S, representing DFA symbols σ_k (see Figure 12.4). In order to derive an interface controller machine ICM from M', these signal associations have now to be translated into either transient conditions or transaction parameter assignments (controller inputs) or interface signal assignments (controller outputs). An interface signal association with a transaction parameter is translated into an interface signal assignment if the direction attributes of this parameter and the interface signal match. Otherwise, it is translated to a transaction parameter assignment.

168   VIRTUAL COMPONENTS DESIGN AND REUSE

Interface signal associations with a constant are translated into interface signal assignments if the direction attribute of the interface signal and the enclosing interface statement match, otherwise a transient condition is generated. Multiple transient conditions are then logically anded.

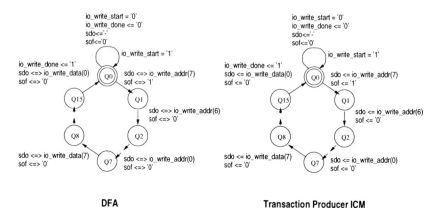

Figure 12.4 DFA and ICM state transition graph for transaction 'io_write'

12.3.2 Substitution of the IP TLI with the Interface Controller

The second step in the interface synthesis process is the replacement of the TLI of the IP core model with an RT specification of the interface controller, so that the final result is a synthesizable VHDL model of the IP core. The top level architecture body of the IP module is enhanced with a set of VHDL processes that form the specifications of the interface controller machines. The declarative part of this architecture is extended with a set of signals that are used to set up the transaction 'send' parameters before the transaction is started and to read the parameters which are 'receive parameters' when the transaction has completed. Furthermore, for each transaction two signals <transaction>_start and <transaction>_done are generated which synchronize the VHDL processes that participate in transactions with the interface transaction controller machines. All transaction join statements are then replaced with the VHDL template shown in Table 12.5, so that the process descriptions become synthesizable. In this template, the loop with the enclosed 'wait' statement, that is triggered to the process clock, ensures that the process execution is blocked until the transaction has completed, which is the behavior of the original transaction join statement. If multiple transactions are mapped to a common subset of interface signals, additional VHDL code for a multiplexer is generated that switches the outputs of the currently active interface machine for this transaction (indicated by the <transaction>_start signal) to the shared communication channel. Finally, in the top level entity description of the IP the interface port statement is replaced with a set of module ports that are derived from the interface signal declarations in the interface specification.

```
-- join interface_port.transaction(param1,param2,...)

<transaction>_send_param1 <= send_param1_ref;
    ...
<transaction>_start <= '1';
while <transaction>_done = '0' loop
   wait until process_clk'event and process_clk='1';
end loop;
<transaction>_start <= '0';

receive_param2_variable_ref:= <transaction>_receive_param2;
```

Table 12.5 VHDL template for replacement

12.4 RESULTS

We implemented the interface synthesis algorithm with about 9000 lines of C++ code in a tool called MODIS[1]. In order to demonstrate the applicability of our method we then specified two different synchronous interfaces for the example IP and generated the corresponding interface implementations with MODIS. The first example defines a 4 port unidirectional serial interface with 2 data lines for serial data input and output and 2 control lines 'start of input/output frame' that indicate the start of an I/O transaction. As a second example we captured the PCI bus protocol specification for single read/write transfers in the IP interface.

In order to evaluate performance, area and customization effort of the interface implementations generated with MODIS in comparison to a hand-coded, optimized interface implementation, we synthesized the VHDL models (MODIS-generated and hand-coded) of our example IP for a XILINX XC4062xl-2 FPGA (using SYNOPSYS Design-Compiler and XILINX PPR) and extracted the corresponding figures for implementation area and maximum clock frequencies.

In Figure 12.5 the implementation areas in CLB's consumed by the IP and the maximum clock frequencies for the two different interfaces are shown, separated into IP core clock and interface clock, which may be distinct. It turned out that the area figures were generally about 15% higher for the MODIS-generated interface implementations. However, the maximum clock frequencies for IP core clock and interface clock as shown in the second diagram are competitive to those of a hand-optimized interface. For the serial interface we even achieved a 25% higher maximum clock rate for the MODIS-generated interface implementation. In this case,

---

1. IP **Mod**ule **I**nterface **S**ynthesizer

the hand-coded implementation contained a shift register and a counter to accumu-

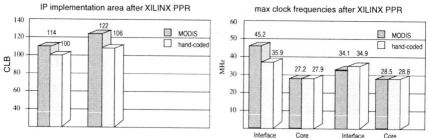

Figure 12.5 IP area and performance for different interfaces

late the number of sent and received bits. Finally, Figure 12.6 compares the customization effort for the IP interface for the two different protocols. While for the hand-coded implementation the user has to actually write and debug VHDL code containing complex state machines, with our approach he only needs to modify the protocol rules in the interface specification and run MODIS, which results in an approximately 400% time saving.

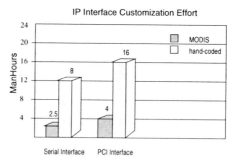

Figure 12.6 Customization Effort for different IP Interfaces

12.5 CONCLUSIONS AND FUTURE WORK

We presented a design for reuse technique for customizable interfaces of Soft IP cores using the VHDL+ extension to VHDL. With our approach, the interface of an IP core can be efficiently customized for a system specific communication scheme which raises the efficiency of IP reuse. Furthermore, in order to adapt the module interface to his own requirements the IP user does not need to have knowledge about implementation details of the IP core, an important aspect regarding IP protection. Our approach is currently limited to synchronous interfaces, however, we are looking into a way to extend our approach to specification and synthesis of asynchronous interfaces using the VHDL+ extension to VHDL.

13 MODELING ASSISTANT - A FLEXIBLE VCM GENERATOR IN VHDL

Andrzej Pulka

Institut of Electronics
Silesian University of Technology
Gliwice, Poland

13.1 INTRODUCTION

In this chapter the Modeling Assistant (MA) - an environment for automated generation of VHDL models of complex digital devices - is shown. The presented system is based on the VITAL Models Generator, an automated standard components generator supplied with AI tools [Pulk97]. Additionally the proposed system contains the library of templates (VITAL library of components).

A special part of the system, the Reuse Engine Module (REM), is responsible for reuse. Each design data about the new object being modeled is transformed to the Generation Entity of the Model (GEM), which is a kind of circuit identifier.

The choice of the VITAL standard of VHDL allows generating sign-off models and using other libraries of components delivered by various vendors. VITAL defines a set of modeling rules, specific constructs and name conventions.

13.2 REUSABILITY OF VHDL MODELS

As mentioned before, design reuse is a crucial task. The problem of software reuse is known from the 60's and since that time many researchers and vendors have proposed different categories of reusable knowledge ([Kiss97], [Math97], [Prei95a], [Prei95b]). In [Kiss97], these types of reusable artifacts grouped into: data reuse (reusing standardized data formats), architecture reuse (reusing standard partitioning), design reuse (reusing common knowledge) and program reuse (reusing code excerpts).

Generally, a reuse approach is characterized by the way in which design artifacts are abstracted, selected, specialized and integrated:

- Abstraction is the manner of circuit representation containing the level of accuracy of the design description.
- Selection describes the process of querying for similarities between the design abstraction and those objects from the existing library which fit into given design.
- Specialization is the process of adopting designed components, which may not exactly match the existing library.
- Integration is the process of combining a collection of selected and specialized artifacts into a complete system.

The reusability can be found in circuit modeling in HDLs on various levels of abstraction - we can distinguish system architecture reuse, internal module reuse, process reuse and library reuse ([Chao95], [Koeg98]).

The proper modeling style may simplify the process of selection and specialization of a given component (simple cell or even a larger subcircuit). The VITAL standard [VITA95] defines a set of requirements that are necessary to develop fast sign-off models in a user friendly way. Those models are compact and easy-to-understand. The reused models of more complex devices have to be constructed as basic VITAL components bound in an entity (VITAL Level 0 [VITA95]). Unfortunately, for mixed A/D devices there is no unified modeling methodology. The VHDL-AMS standard is not popular and it offers a wide range of interpretation for design reuse. Some subsets of the mixed systems discussed in ([Dabr98a], [Dabr98b]) can be modeled with the PWL (Piece-Wise Linear) technique, thus enabling a digital VHDL environment. The continuous analog waveforms are divided into discrete time events and linear segments. If the PWL technique is assumed as a methodology of modeling analog parts of the system, the proposed reuse methodology can be extended to a class of mixed A/D systems [Dabr98b].

13.3 DATA MODEL

Figure 13.1 presents the architecture of the proposed system for reuse: the Modeling Assistant (MA). Two main parts of the system can be distinguished: the NVMG (Nonmonotonic VITAL Models Generator) and the REM (Reuse Engine Module). The NVMG generator is responsible for generation of new parts of the system that can be reused. The abilities and features of this automated generator are described in [Pulk97]. The NVMG contains a set of several intelligent tools supporting the process of VHDL (VITAL) model generation. Each new generated element (component model) is added to a database and after that it owns a "shadow" in form of listframes [Pulk97] which offers a frame-like knowledge representation of the device (Table 13.1). This redundant storage of information simplifies modifications of models on the lower level of knowledge representation.

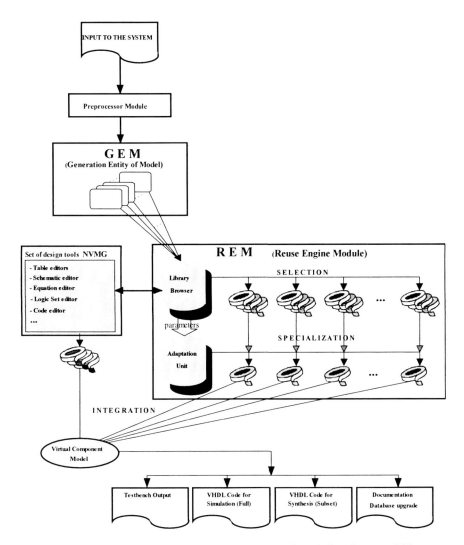

Figure 13.1 Modeling Assistant - the system of modeling for reusability

The REM module plays the role of a reuse supervisor in the system and it has two special built-in tools: a browser and an adapter. The process of modeling can be divided into several characteristic steps. Firstly, the initial data is transmitted into the system via a GUI (Figure 13.2) which transforms it into the hierarchical form of the GEM (Generation Entity of the Model). The GEM is a very important part of the data because it constitutes a kind of model identifier (label) and the system tools communicate with each other via the GEM. The main features and examples of information stored in the Generation Entity of Model are addressed in the following section.

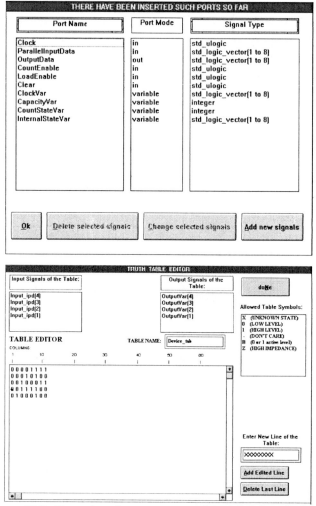

Figure 13.2 Examples of menu system of the Modeling Assistant

In the next step, the REM is executed and a browser queries for a proper element in the library of templates (structures, modules, architectures, processes etc.). In parallel, it tries to find elements which match the data stored in the GEM. The GEM is hierarchically checked (frame by frame) and if the result of a search in the library fails the NVMG generator (Figure 13.3) [Pulk97] supports the modeling process. The adoption unit is a further tool of the REM, which reworks the selected elements of the library and adds the actual parameters extracted from the GEM.

MODELING ASSISTANT - A FLEXIBLE VCM GENERATOR IN VHDL

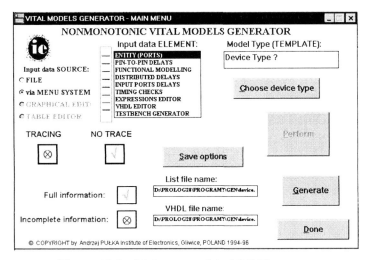

Figure 13.3 Main menu of the NVMG generator

The elements proposed by the REM are integrated with those which are generated by the NVMG. Finally, the Virtual Component Model (VCM) generation is completed. The generated VCM contains testbenches, proper code subsets for synthesis and list-frames for the library storage.

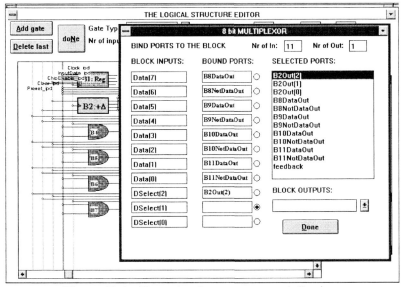

Figure 13.4 Example of component specialization

176   VIRTUAL COMPONENTS DESIGN AND REUSE

The generation scheme proposed meets the main (formulated above) requirements of reusability: the GEM is a kind of knowledge abstraction, the library browser (REM) performs the selection, the adaptation unit (REM) specializes selected modules (Figure 13.4) and the integration of the VCM is performed outside the NVMG and REM modules.

13.4 GENERATION ENTITY OF MODEL (GEM)

The GEM simplifies the reusability of models. The library of existing models (templates) together with the components generated by NVMG generator tools (schematic editors, table editors etc.) have "shadows" in a form of listframes (Table 13.1). These listframes allow to handle and transform the information into an intermediate form. This property is also useful during the selection and specialization procedure.

The information stored in the GEM has a hierarchical and classified form that offers a lot of flexibility to the communication between GEM (Table 13.2) and listframes (Table 13.1) (library). When it is necessary to generate new components, e.g., if the selection process fails, additional information that is stored in the GEM will be useful for the nonmonotonic inference engine during the new component generation [Pulk97].

As it can be appreciated in the fragment of GEM is referred above, its syntax is compatible with the LPA-Prolog (Prolog clauses) - the system implementation language [Pulk97].

```
listframe([model,1,processor,'MCS51']).
listframe([model,2,decoder,'BitGrayDecoder4']).
listframe([microcontroller,2,'MCS51']).
listframe([port,1,'XTAL1','in','std_logic']).
listframe([port,1,'RESET','in','std_logic']).
listframe([port,1,'P3','inout','std_logic_vector(0 to 7)']).
listframe([function,2,decoder,['InputData','ChipEnable',
          'OutputData']]).
listframe([function,2,truthtable,'BitGrayDec4_Tab',
          ['InputA','ChipEnable','OutputData']]).
listframe([function,'BitGrayDec4_Tab',2,1])
listframe([function,'BitGrayDec4_Tab',
          '('1' '-' '-' '-' '-' 'Z' 'Z' 'Z' 'Z')'])
listframe([function,'BitGrayDec4_Tab',
          '('0' '1' '0' '0' '0' '1' '1' '1' '1')'])
listframe([function,'Butt-01',lowpass,1])
```

Table 13.1 Fragment of the library contents - examples of listframes.

```
gem([design_name,"AM2901"]).
gem([design_name,"m-8051"]).
gem([design_name,"dsp"]).
gem([design_type,"m-8051",processor]).
gem([design_type,"dsp",processor]).
gem([design_class,"m-8051",
     [microcontroller,buffered,programmable,...]).
gem([design_class,"dsp",[signalprocessor,floatingpoint,...]).
gem([design_ports,"AM2901","AddrA","AddrB","Clock",..]).
gem([design_module,"AM2901","ALU-01"]).
gem([design_module,"TrackHold001","AMPL-01"]).
gem([design_module,"TrackHold001","INER-01"]).
gem([module_type,"ALU-01",arithmeti]).
gem([module_type,"TrackHold001","AMPL-01"]).
gem([module_type,"TrackHold001","INER-01"]).
```

Table 13.2 Examples of the GEM syntax.

13.5 EXPERIMENTS AND RESULTS

The system implemented in LPA Prolog has been tested with several types of modules and architectures - an example of standard components library has been generated (Table 13.3 presents a fragment of them).

```
entity Am2901 is
generic(tipd_AdrA:
       VITALDelayArrayType:=(0.1ns,0.1ns,0.1ns,0.1ns);
       tipd_CLK: VITALDelayType01   := (0.02ns, 0.03ns)
(...)  tsetup_AdrA_CLK: VITALDelayArrayType   := (1.2ns,
1.2ns, 1.2ns);(...));
   port(AdrA, AdrB:      in STD_LOGIC_VECTOR (3 downto 0);
       Y:                out STD_LOGIC_VECTOR(3 downto 0);
       D:                in STD_LOGIC_VECTOR (3 downto 0);
       INSTR:            in STD_LOGIC_VECTOR (8 downto 0);
       CLK, Oebar:       in STD_LOGIC;
       Cn, Cn4, OVR:     in STD_LOGIC;
       Gbar,Pbar,Z,F3:   out STD_LOGIC;
       Ram0,Ram3:        inout STD_LOGIC) ;
ATTRIBUTE VITAL_Level0 OF Am2901 : ENTITY IS TRUE;
end Am2901;
```

Table 13.3 Fragment of the VITAL Level 0 entity of the AMD2901

The idea of GEM and concept of reusability reduces the number of NVMG tools invocations (Figure 13.3) and consequently the generation times. The results show that the number of modified lines does not exceed 20 percent of the whole VITAL code.

Furthermore, another problem concerns the reusability of testbenches since the process of automated generation of exhausted testbenches is often very difficult and time consuming. So it is important to make the system able to reuse whole (or even parts) of the testbenches or to expand the existed testbenches. The MA system is supplied with the mechanism that produces automatically testbenches for the devices with regular structures. Figure 13.5 presents results for the typical Gray coder/decoder.

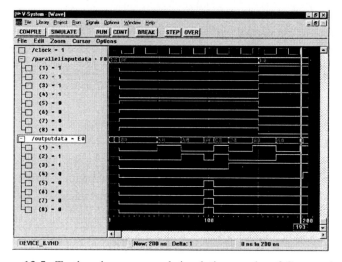

Figure 13.5 Testbenches reuse and simulation results of Gray coder

For the "pure" digital blocks, the VITAL standard offers a lot of primitives – functions and procedures - that allow handling data as well as timing in a given model.

The mixed A/D systems, however, require a special analog package with sets of conversion functions, types definitions and the PWL package that contains functions responsible for the signal approximation [Dabr98b]. The example in Table 13.4 presents the fragment of the track-hold module description and the example of waveforms of this model simulation obtained with VSystem (Model Technology) simulator are illustrated below (Figure 13.6).

```
ENTITY track_hold IS
GENERIC Pmax: real:= 0.01; Tau:time:= 100 ns);
PORT
      (TH:      IN std_logic;
      Tend:     IN time;
      Rin:      IN real;
      Xpwl:     IN voltage;
      Tendout:  INOUT time;
      Rout:     INOUT real;
      Uout:     INOUT voltage);
END track_hold;
ARCHITECTURE only OF track_hold IS
   -- SIGNAL declarations
   (...)
   PROCESS(TH,Rin,Tend,SelfStrobe)
      -- VARIABLE declarations
   BEGIN
      IF Initial THEN
      -- initializations;
      ELSE
            -- actual PWL segment calculations
      END IF;
      -- PWL approximation
         IF NOT(Start) THEN
            IF(TH = '0') THEN
            -- variable calculations
            Tstep    := real_to_time(T1);
            Tendout <= Tstep + NOW;
            Uout    <= Xnext;
            (...)
            ELSIF(TH = '1') THEN
            -- variable calculations
            Tstep    := real_to_time(T1);
            Tendout <= Tstep + NOW;
            Uout    <= Xactual;
            (...)
            Rout   <= 0.0;
            END IF;
         END IF;
      (...)
      IF(Activated) THEN
         SelfStrobe <= NOT(SelfStrobe) AFTER Tstep;
      END IF;
   END PROCESS; END only;
```

Table 13.4 Fragment of entity description of the track-hold model.

180   VIRTUAL COMPONENTS DESIGN AND REUSE

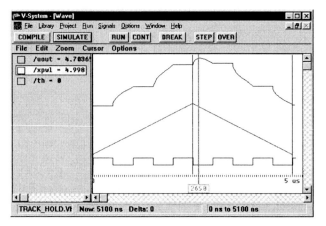

Figure 13.6 Simulation results of track-hold model

The next example recalls the bi-quad structure presented in [Dabr98b] composed from reused and independent blocks: integrator and inertial block. It shows the fourth row low pass filter, which consists of two bi-quad sections (Figure 13.7and). This introduces another feature of this methodology hierarchical reuse - each of the reused bi-quad sections contains reused basic blocks.

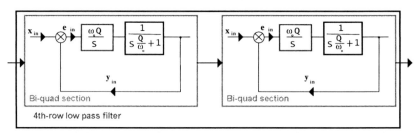

Figure 13.7 Example of 4th-row low-pass filter with 2 reused blocks

13.6 SUMMARY

The system proposed can be integrated into high-level synthesis tools, and thanks to the VITAL standard of components and VITAL Level 0 [VITA95] compliant entity of the top-level model (Table 13.3) it can be useful for modeling and post-layout simulation while accompanied with proper SDF files [OVI95]. The VITAL standard allows also to use a library of templates containing models developed by other engineers.

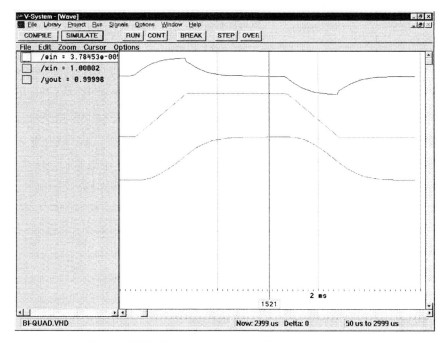

Figure 13.8 Simulation results of track-hold model

The approach proposed seems to be very suitable for digital signal processors, where reusability can drastically reduce the system model generation time.

The prototype has been tested with several A/D systems and the results are promising if a unified PWL algorithm [Dabr98a] is employed for analog blocks (Figure 13.6, Figure 13.7). The presented approach simplifies the reuse process of basic blocks and in many cases only a few generic parameters have to be changed (Table 13.5). However, the automation of the generation of such A/D systems remains difficult because a set of a sophisticated rules have to be implemented in the inference engine and continuos interaction is needed.

A remaining task for analog and mixed signal models developers is to establish a standard (like VITAL) for VHDL-AMS, in order to make fast progress in behavioral modeling of analog and mixed A/D circuits. In principle, the VITAL standard requires a special set of types for signals, variables and timing procedures. And for mixed A/D models this model has to operate with the tuples of signals defining PWL segments [Dabr98b] for instance: slope, time and initial value.

dev. name	device type	full gen. time[s]	gen. time [s]	nr of lines	mod. lines	Vital code [kB]
12gray	12bit Gray decoder	9.42	6.2	800	5	40.2
14add	16bit adder	6-7	3-3.6	2250-2400	11-125	111-127.4
12cnt	12bit counter	22.2	15.7	1061	10-330	54.1
4par	4bit parity control	5.7	3.9	150	3-26	6.7
8reg	8bit shift register	11.3	8.9	750	86-140	31.5
4bsreg	4bit bidir. shift reg.	6.4	4.3	300	17-71	11.7
84alu	8x4 alu	28.2	23.2	1400-2100	150-560	28-37
82pld	8inpx4out prog. PLD	31.3	22.1	800-1100	102-400	27.5-28
81ram	8x1bit RAM	42.5	34.3	2500-3200	256-800	186.5

Table 13.5 Examples of generated library of VITAL templates

The experiments proved the benefits of this methodology, but it has still some unsolved parts, e.g., Virtual Component Models (especially for more sophisticated devices) have to be corrected (adapted) manually because the automated generator is not able to cover all possible solutions and user interactions that are necessary, even if the nonmonotonic mechanism is called [Pulk97].

14
REUSING IPs TO IMPLEMENT A SPARC® SOC

Serafín Olcoz, Alfredo Gutiérrez and Denis Navarro*

SIDSA
Semiconductor Design Solutions
Madrid, Spain

*Universidad de Zaragoza
Departamento de Ingenieria Electronica y Comunicaciones
Zaragoza, Spain

Abstract

We present the gained experience dealing with the implementation of a SOC, reusing existing soft-cores. The application of a hardware-software virtual prototyping (co-simulation) approach used to functionally verify the system is also presented. This is specifically tackling the advantages and disadvantages of such a approach. The preparation of the deliverables-list needed to offer the soft and hard-cores as part of the semiconductor IP-products catalog. The communication description between the co-simulation and the soft-core management environment is presented in a separate section.

14.1 INTRODUCTION

IP hard, firm and soft-cores are occupying significant portions of new designs, particularly in the fast-growing consumers electronics market where a strong application demand the driving factor. IPs provide off-the-self implementation expertise, and a means of vastly improving gates/day productivity.

The compelling feature of IPs is that they are correct-by-construction, allowing for the re-deployment of engineering resources to other critical aspects of the design. Using IP solutions will provide a dramatic time reduction in design cycle. Reuse is becoming as necessity as gate counts rise to 10 million gates complexities of a System-on-Chip (SOC) design. Time to market pressure is driving the IP revolution; designers have no longer the luxury of a 12 months design cycle to deploy systems. Figure 14.1 describes the IP revolution as the percentage of new design content in a system as a function of time [Mcle97].

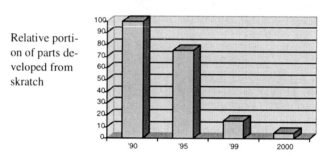

Figure 14.1 Decreasing percentage of new design content in a SOC

In the early part of this decade, conventional ASIC design were less than 500k gates, and most of these gates were developed by hand via HDLs-based design techniques. The major emphasis was on minimizing area consumed and maximizing performance provided; this has been achieved via the aforementioned HDL-based hand crafting and additional gate-level optimization.

With the SOC trend, in which designs will approach 10M gates, the conventional ASIC design methodology has to be modified to accommodate the new emerging emphasis: minimum design time and optimization at the system-level. Today's designs cannot be managed at the gate-level. The leverage necessary to achieve SOC will be supported by IP cores application.

Figure 14.1 predicts a dramatic impact of IPs in future SOC designs. In order to meet TTM objectives, larger and larger portions of the system will consist of IPs, either purchased on the open market or developed internally.

The next section introduces the work done to develop a SPARC μ-Processor, from HDL to ASIC implementation, and finally, to design an IP core. Section 14.3 introduces a similar approach carried out to develop some μ-Processor peripherals. Section 14.4 presents a SOC development based on reusing these IPs, and Section 14.5 presents current results obtained in the CoMES project. The contribution will be finished by a conclusion and an outlook.

14.2 SPARC μPROCESSOR IP

The company TGI (http://www.tgi.es) implemented a SPARC Integer Unit on silicon, starting with VHDL to Silicon[1], between middle of 1992 and beginning of

1994. The next subsection introduces the SPARC architecture and the following subsections present issues about its silicon implementation and further transformation into a soft-core.

14.2.1 The SPARC®Processor Architecture

The Scalable Processor ARChitecture (SPARC®) was defined by Sun Microsystems Inc. between 1984 and 1987. A key part of the SPARC approach is its foundation on open systems, i.e. the availability of compatible technology from multiple suppliers. SPARC has reached a level of openness, and heretofore, unimagined in the computer industry. This open system philosophy does not only affect to general-purpose applications but it provides a good candidate for microprocessor based embedded systems.

Currently, the evolution, standardization and OS environment of this microprocessor architecture is directed by SPARC International Inc., which is an independent and non-profit consortium founded in 1989. Its members are software, hardware and microprocessor vendors, distributors and users committed to the SPARC architecture. The aim is to influence the evolution by open standards. SI documents this evolution through two primary vehicles: The SPARC Architecture Manual [Spar91] and the SPARC Compliance Definition (SCD, [Spar92]). These two documents provide vendors of chips, systems, applications, and add-in/add-on hardware information about SI standards, and how to ensure binary level compatibility with all other SPARC products. Moreover SPARC was proposed as an IEEE standard (PAR. 1754).

The SPARC-based system is a processor based system composed of the following design elements: a processor (a SPARC processor logically comprises an Integer Unit (IU), a Floating Point Unit (FPU), and an optional coprocessor (CP) that has not been considered here), and a storage system composed of a Memory Management Unit (MMU) (following the reference provided by SPARC although it is not part of the architecture definition), and a memory unit. SPARC assumes a linear, 32-bit virtual address space for user application software. In this work, the data bus size is based on a 32-bit data bus. An 8-bit address space identifier field (ASI) can be used to manage system resources. This allows the system address space to be divided into 256 separate 4 GByte areas. The storage system could be extended in order to have a more complex hierarchical memory (e.g., containing data and instruction cache memories).

A feature that distinguishes SPARC from other RISC architectures is that these implementations are using overlapping register windows [Kate83]. They reduce the number of loads and stores as well as simplifying the compiler register allocation algorithms. In this schema, results placed in registers need not to be explicitly saved

---

1. Partially funded by ESPRIT-GAME Program, project IdeAS (Implementación de la Arquitectura SPARC) and in cooperation with University of Zaragoza.

or stored during procedure calls. The scalability of SPARC refers to its flexible integration of cache memories, MMU, FPU, multiprocessor, bus size and the number of CPU registers (from 2 to 32 register windows, each window has 24 working registers plus 8 global registers) that can be implemented in various versions. This flexibility allows implementations at a suitable price/performance.

Register windows can be managed to tailor the operation of SPARC to specific needs (e.g., real time applications, requiring fast context switching and deterministic behavior,). Moreover, the architecture has been enhanced for embedded systems, by adding several extensions (such as bit field operations or simplified MMUs) that are contained in the V8E architecture extension [Spar94]. Among many implementations already existing in the market, we chose the Cypress CY7C601 [Cypr90], for developing the pin-out compatible demonstrator chip. This decision also implies a four-stage pipeline.

14.2.2 From VHDL to Silicon

We want to remark the main barriers had to be removed in order to produce an ASIC (ASIC, Application Specific Integrated Circuit). Besides that, a CAD flow has been defined and a silicon foundry (ES2) design kit for standard cell (0.7 µm, 2MTL technology) has been applied (see Figure 14.2),

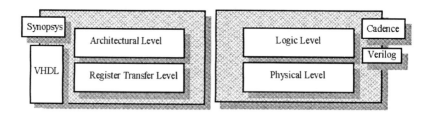

Figure 14.2 IDeAS project CAD Flow

The following modeling approach consists of describing a complete behavioral model in VHDL. Complete means that the design can be executed as a real SPARC application software and not only like a VHDL description of the IU and its corresponding abstract test benches. The board can be considered as a functionally detailed test bench of the IU. However, at this stage of the design the same model of the board could be used to design and manufacture the FPU instead of the proposed IU design.

Obtaining a self contained board model, we need to add some auxiliary units to the SPARC processor and the storage system: an initialization unit (reset and software loader) that loads the initial state of the different units and it must load software programs in the memory. The interrupt unit models the asynchronous interaction with the external world and a clock unit.

The total size of the board model is about 20.000 lines of code (6.000 lines correspond to the behavioral model of the SPARC IU). The simulation performance of this board depends on the commercial simulator and hardware platform used to measure it. In a SPARCStation 10, it was possible to execute between 250 and 400 SPARC instructions per second by using Synopsys and Cadence simulators in 1994.

The application software is clearly part of the board model and it can be generated by using a C programming environment that is able to produce SPARC machine code (e.g., the C compiler provided in each Sun SPARCStation, see Figure 14.3-A).

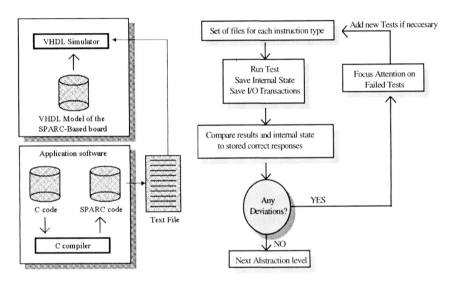

Figure 14.3 (A) HW-SW Co-simulation. (B) Regression test method used

After producing the machine code, the application software is stored into a text file that can be interpreted by a unit of the VHDL model. This unit, called software loader, reads the machine code and performs the loading of the software into the model that represents the memory. With this support a possibility was provided to model an instruction set accurate VHDL behavioral description that can be used as a platform for hardware-software cosimulation. The test strategy followed to validate this VHDL model is based on its capability of executing software (see Figure 14.3 (B)). A more detailed description can be found at [Olco99].

Using RTL synthesis and place&route tools, the design was refined, and after signing simulation a design tape was submitted to ES2. The silicon implementation of the SPARC Integer Unit was successful and TGI licensed it to ES2. It can be relicensed as a firm or a hard-core. The firm-core consists of placement and boundary data, clock/power, place and route scripts, floorplanner scripts, clock tree analysis and extracted estimated interconnect parasitics for back-annotation into simulation.

The hard-core consists of a layout database, DRC/LVS files, manufacturing test vectors and an emulation or prototype board.

14.2.3 From ASIC implementation to IP

In order to guarantee that SPARC products remain binary compatible, SI's compatibility and compliance committee established the SPARC compliance definition (SCD), that consists of a set of specifications of common interfaces. By developing systems and software that fulfil the SCD, SPARC system vendors and application developers ensure that their products are binary compatible and interoperable.

SI established SCD testing suites at both system and application levels. Upon successful completion of the compliance testing suites, vendors may label their products with the SPARC term to denote verified SPARC binary compatibility. It was proposed that HDL descriptions of SPARC architecture components should be a product by themselves and not just a semiconductor part or the hardware and software built with them. In order to do so, we proposed to follow our co-simulation approach. This performs the same kind of certification process on HDL descriptions that was carried out with semiconductor parts on ATEs (Automatic Test Equipment). This implies to implement a new kind of licensing agreement to open SPARC technology for IP reuse.

In the VHDL-Verilog co-simulation environment the regression test was passed again by using SPARC binary coded test suites for input [Usse93]. SI validated the existing implementations of the SPARC architecture.

The same process was performed with the HDL descriptions corresponding to all abstraction levels involved. The validation of the gate level description was used to certify the demonstrator chip produced by ES2. The soft-core consists of a RTL HDL testbench (co-simulation board) and vectors, synthesis and test insertion scripts, generic library gate-level netlist and documentation. This work has been finished in 1994, before the IP revolution started.

14.3 μPROCESSOR PERIPHERALS FOR EMBEDDED SYSTEMS

TGI was also participating in SMILE (SPARC Macrocell and Interface Library Elements project) project to develop μ-Processor peripherals for embedded systems. These peripherals (two general purpose USARTs with internal FIFOs, an interrupt controller) were developed according to the SPARC Embedded Architecture Definition, [Spar94].

These peripherals were also implemented in silicon using MATRA MHS 1 μm MC/TC gate-array technology. It was realized as an MCM (Multi Chip Module) together with the SPARC microprocessor, developed by MATRA and an image processing macrocell, from Sussex University. Everything was connected via a PI-Bus (Peripheral Interface Bus, [Pibu94]).

The initial trading IP business started in the area of peripherals, and so the peripheral bus (PI-Bus) was used in the SMILE project. Even without formal standardization, many companies adopted it.

TGI's objective was to use these peripherals together with their own SPARC IU, in order to build a macrocell library for joining the emerging IP business. However, this market emerged slower than expected by OMI. Even now, three years after IP explosion, its business model is not clear at all.

14.4 A μCONTROLLER FOR A SCALABLE REMOTE TERMINAL UNIT

TGI started a new OMI project to develop a SOC based on the SPARC architecture. The objective of the project was to develop a scalable SCADA (Supervisory Control and Data Acquisition) remote terminal unit (RTU) for telecommunications and industrial control applications.

The family of DISEL's RTUs includes intelligent units that allows to remotely monitor and control the equipment. The RTUs are designed modular and they contain highly intelligent processing capabilities. The RTUs perform routine tasks; gather information, process, calculate, evaluate and convert the information; and they transmit only pertinent data to the master (report-by-exception), in a hierarchical and scalable way. By performing calculations and repetitious jobs, the RTUs free up the master unit for other vital tasks while decreasing the amount of useless information.

If communications between the master and an RTU is lost, the intelligent remote unit continues to perform its pre-programmed functions and accumulate status changes. It stores the data until communication is restored and then it feeds it back to the master, therefore, no data are lost. RTUs are modular in design. This allows convenient maintenance and cost-effective extension possibilities. And DISEL customizes each configuration based on customer needs and site requirements. The objective of DISEL was to design a scalable RTU, based on ECU SOC, to be used at any level of the hierarchical network RTU, from data acquisition level to the data center level, where, even SPARC-based workstations can be used to complete the chain. This approach also unifies the software development and maintenance environments.

ECU SOC is composed of a new version of the integer unit, coming from reusing the soft-core developed in the IdeAS project, together with some microprocessor peripheral soft-cores, coming from SMILE project. It is designed to fulfill and extend features specified by the SPARC"-V.8e architecture definition for embedded systems, but it is not restricted to interface this kind of CPU and data/instruction cache memories.

In order to develop this chip, the bus interface was simplified. System support functions were built in for minimizing the amount of glue logic required in the external system. It includes system-configuration registers (six different chip selects, wait state generation), a timer for generating refresh request, and a same-page detection mechanism. It contains a user/supervisor space mapping and it is a no-multiplexed address/data bus.

The peripherals were develop as soft-cores, and therefore, reusing them was very easy and it just addressed technology retargeting issues. The SOC was implemented using standard cells in CMOS, 0.7 µm 2MTL ATMEL-ES2 technology (Figure 14.4).

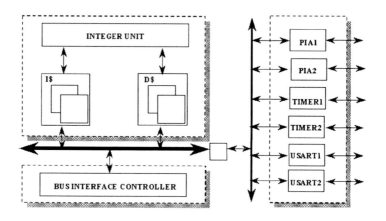

Figure 14.4 Block diagram of ECU SOC.

The CAD flow and test strategy were taken from the IDeAS project. DISEL provided the corresponding firmware to carry out software and system level tests (see Figure 14.5). These tests support hardware debugging, e.g., to test combinations of instructions when debugging real application software (including a commercial RTOS). The tests were based on the HDL version of the microprocessor, including cache memory management tests. Verifying this SOC was much more complex than verifying the IDeAS µ-Processor. We had to improve the co-simulation environment used for validation purposes [Olco95a]. So, we developed a hardware-software co-debugging environment [Olco95b].

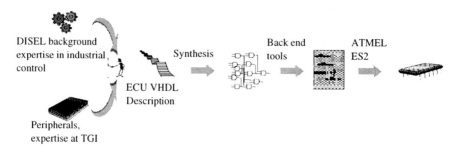

Figure 14.5 ECU SOC design process

At operating system intercommunication process level. EDS, links the VHDL simulator of Synopsys with the GDB software debugging tool of GNU. This is performed via a hardware debug and profile tool (HDPT) (see Figure 14.6). EDS allows to debug application software and hardware. The advantage is that system, software and hardware engineers can share one platform. The disadvantage is the reduced performance, which is insufficient for software engineers and too poor for system engineers.

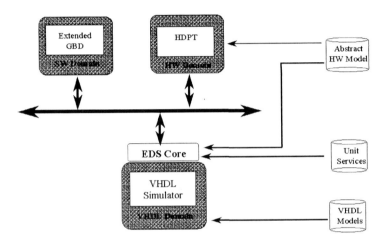

Figure 14.6 EDS architecture

However, software and system engineers were able to develop and test the BSP (board support package) on the HDL simulation models, as a part of the board porting of the commercial RTOS. Application drivers were also developed and used as part of the system level debugging activities carried out on the EDS. The VHDL-ICE design management environment allows to trace dependencies among softcores and to navigate through its source code. Hierarchy and simulation process models can be applied for inspection and evaluation guidelines.

14.5 COMES: CO-DESIGN METHODOLOGY FOR EMBEDDED SYSTEMS

In the CoMES project Co-Design Methodology for Embedded Systems) that is based on MCSE methodology [Calv93], SIDSA's objective is to establish a link between the EDS and the VHDL-ICE Design Management Environment, that has been developed in TOMI (Tools for OMI) and REQUEST (REusability and QUality ESTimation) OMI projects ([Olco98a], [Olco98b], [Olco98c]).

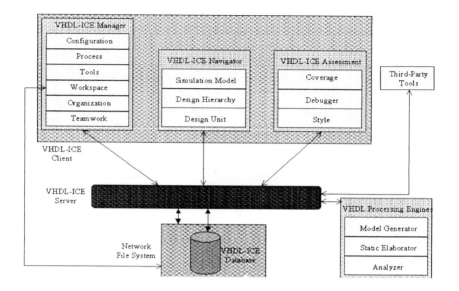

Figure 14.7 VHDL-ICE architecture

VHDL-ICE is a VHDL Integrated Common Environment, suitable for soft-core reuse (see Figure 14.7). The VHDL-ICE client/server architecture supports and inter-operates across multiple platforms (Windows NT, UNIX). The base middleware is based on sockets and TCP/IP. It allows expandable and configured installations over local or wide area networks, i.e., Intranets and the Internet, without forcing VHDL developers to change their development tools or the way they work.

To establish a link between EDS and VHDL-ICE, it has to be mandatory to update the EDS environment from SunOS to SOLARIS and taking into account the differences in inter-process communications between both operating systems.

The following approach has been very pragmatic and it is based on exchanging files between both environments (see Figure 14.8). This approach takes advantage of the hierarchical import/export capabilities offered by VHDL-ICE. It transfers hierarchical and interdependent libraries corresponding to a SOC design data base.

14.6 CONCLUSION AND FUTURE WORK

We have presented the gained experience in several projects dealing with the implementation of a SOC based on the SPARC architecture by reusing existing soft-cores. The preparation of the deliverables-list needed to deal with soft, firm and hard cores. It is part of the semiconductor IP-products catalog and the certification process carried out by SPARC International. All cited cores are currently available through the RAPID catalog (http://www.rapid.org) and the Design & Reuse yellow pages (http://www.design-reuse.com).

The application of a hardware-software co-simulation approach used to functionally verify the system has also been presented, specifically tackling the advantages and disadvantages of such an approach. The communication developed between the co-simulation and design management environments has also been presented. In the future, it is planned to extend the development environment by means of co-emulation systems, such as the HSDT100 and new HSDT200 development boards.

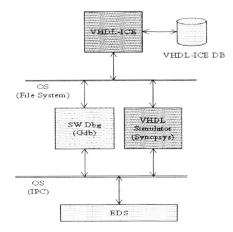

Figure 14.8 EDS / VHDL-ICE communication

15 HARDWWWIRED: USING THE WEB AS REPOSITORY OF VHDL COMPONENTS

Adriano Sarmento, Jorge Fernandes and Edna Barros

Centro de Informática, UFPE, Brazil

15.1 INTRODUCTION

The rapid technological development of the last thirty years has provided means for the design and fabrication of complex digital systems [Mich92]. Systems that in the past took several months and people to be designed can nowadays be made in a few weeks by a single person. These improvements in design and fabrication of digital systems were mostly due to development of design methodologies and CAD tools for automated design. An important support in the design of digital systems is the reusability of components, i.e. to design new components from existing ones. The main benefit of this feature is quite clear as it reduces enormously the design time by not requiring from designers to create basic and generic components every time they start a new project. As digital systems become more and more complex, reusable components play an essential role. Of course, reusable components would not bring so many benefits if there was not any common place to store them so that people could easily share these components, this is why repositories of reusable components are also important. However, one major problem in creating reusable components and so these repositories is that the design process usually requires the use of several CAD tools which accept different description languages and therefore make harder the components reusability.

The incompatibility among the different tools led to the development of a hardware description language called VHDL (Very High Scale Integrated Circuits Hardware Description Language) which should be employed as a standard format for tool interoperability and therefore would lead to reusable components design.

Even though this reduced the incompatibility, it was not enough because different tools accept different subsets of VHDL.

In the late eighties an important effort for tool integration was done by the introduction of frameworks [Wagn94]. Frameworks are platforms for building open and integrated electronic design environments. Although frameworks became very popular in the CAD tool industry, the incompatibility in using them in different platforms was an important restricting factor. More recently the academic community has been focusing their efforts in developing design environments that integrate several tools and create repositories of reusable components using the World Wide Web as platform. The Web is nowadays one of the most popular Internet services. Web pages can not only contain documents but also software applications that are available for all Web users. These features make very attractive the idea of using the Web as a common environment for digital systems design. Examples of these environments are Cave [Indr97] which focuses on specific parts of the design process such as layout and schematic editing, and Weld [Weld], which is centered in workflow and project management. A good design environment in the Web should have a common data format which can not only carry all the information needed by the tools that are integrated in the environment but also can be exchanged easily in the Web. This environment should also have policies that enable tools to visualize the data in a simple and a secure way (controlling non-authorized access). As this data format is defined, tools and users can share it in an efficient way. Another issue that is important is what benefits the Web can bring to such design environments. The main benefits from using the Web in design environments are:

- Sharing and low cost maintenance of resources - Usually the design process of a digital system involves several people who share the same resources. A single copy of a CAD tool or a hardware component can be shared among several people. This helps creating collaborative work environments. Besides the maintenance cost gets lower as the number of copies of the same tool decreases.

- Universal access - Usually CAD tools demand great processing and memory resources and also require specific platforms and operational systems making harder their use in portable systems. The Web solves these problems by giving support to a distributed environment where the bulk of the processing may be done at the server and also by enabling access to multimedia resources to users at any point of the Web, given that the resources that pass through the Web (Java, HTML, images, etc.) are platform independent.

The main goal of this work is to provide an environment for the design of VHDL digital systems in the Web called HardWWWired. This platform should not only integrate several CAD tools but also manage a repository of VHDL components.

The chapter structure is: Section 15.2 describes the architecture of HardWW-Wired; Section 15.3 explains how and why VHDL descriptions are mapped to Java objects; Section 15.4 discusses the security policies employed in this work;

Section 15.5 shows a case study of digital system design using HardWWWired; and Section 15.6 brings the conclusions and future works of this project.

15.2 HARDWWWIRED: THE ARCHITECTURE

The architecture of HardWWWired is illustrated in Figure 15.1. It follows the traditional client-server model where the set of Web tools is located at the user machine and the Web Resources Repository is in a remote computer with greater processing and memory resources. These components are integrated in the Web through an Object-Oriented Web Server called Jigsaw [Bair96]. Jigsaw internally represents each resource as an object that can have its state and functionality changed in a dynamic and persistent way. Jigsaw manages the Web Resources Repository which contains VHDL project units, CAD tools and filters. These filters couple transparently to Project Units and CAD tools and handle each HTTP request to a resource and each HTTP resource response to a Web tool. They play an important role as they can offer many views of the same project to the user or a CAD tool. What is meant by views is that the user may enter a VHDL textual description of a circuit and wish to have a schematic or hyper-textual representation of this circuit, so these filters act as translators of one representation model of the circuit to other ones. Another important feature of filters is that they may control resource access.

Users may access these resources by using the Web tools which are Java applets that represent HardWWWired front-end in the Web. They execute inside a Web browser that supports Java applets and communicate to the server through the HTTP protocol [Fiel]. They represent the interface between users and the environment and offer functionality such as schematic edition, textual edition, simulation, etc.

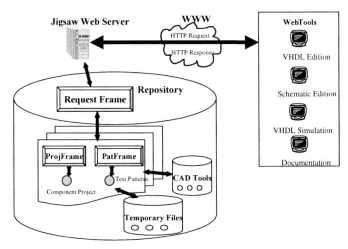

Figure 15.1 HardWWWired Architecture

The communication flow is also illustrated in Figure 15.1. When a Web tool makes an HTTP request to the Jigsaw server, this request is analyzed by the *Request Frame*, a kind of filter located at the server, and depending on the type of request it reaches one of the resources (they are represented in Figure 15.1 by *ComponentProject*, *TestPatterns*, etc.) in the repository. Each resource has filters (they are represented by *ProjFrame*, *PatFrame*, etc.) that can perform many actions such as translating from one data representation to another one, processing data using CAD tools located at the server, or building temporary files needed for some action to be executed. After the filters process the request a response is sent back to the client (Web tools). This response may be only a message indicating that the request was performed successfully as in a store request, or may be a resource requested by the client as in a retrieval action, etc.

This architecture not only enables access to hardware components through the HardWWWired user interface, but also through any HTML browser even if it does not support Java. This is so because every component has a URL attached to it. So users can type in the browser the component URL and access the VHDL description of a component. Of course, this kind of access does not allow users to access private components, only public ones.

15.3 MAPPING VHDL DESCRIPTIONS INTO JAVA OBJECTS

In order to permit a VHDL description to have distinct views each VHDL circuit is represented as Java objects in the HardWWWired repository.

VHDL [IEEE93] is a hardware description language that became an IEEE standard in 1987. It supports descriptions in behavioral and structural domains and several abstraction levels, from the logic level to the system level. This extensive capacity of representativeness has made VHDL a huge language and this is the reason why most of the CAD tools available supports only a subset of VHDL. The basic components of a VHDL description are: **(a) entity** which describes the external characteristics of the circuit, i.e. the circuit's name and its ports, and **(b) architecture** which describes the internal characteristics of the circuit, i.e. the actual implementation of the circuit.

Java [Arno96] is a general-purpose language designed at Sun Microsystems Laboratories that is object-oriented, platform-independent and whose programs can migrate between server and client and execute in a secure manner. Java applications can be executed inside a Web Page (applets), requiring from the user neither great processing resources nor any specific operating system.

In order to implement the mapping of VHDL descriptions in Java, a set of Java classes were defined, each one representing a VHDL construction. When the mapping is performed Java objects are composed hierarchically where each non terminal VHDL BNF element is mapped into an object, and its attributes represent other elements that compose its BNF (these attributes work as links to other objects). This process continues until a terminal element is found. This element then has only Java basic types attributes such as: Boolean, Strings, etc. An example of this mapping is illustrated in Figure 15.2, where the entity *comparador* is mapped into Java objects. The left part of this figure has the VHDL BNF for an entity and the right part has its related objects. It can be noticed that some BNF elements are optional in an entity declaration, when they are not present in an entity the attributes related to them are set to null.

A central feature of HardWWWired is the VHDL 1076 compliant compiler developed for this work using JavaCC [Sunt98] compiler generator. This compiler maps VHDL descriptions into Java objects. During the syntactic analysis, the compiler builds a syntax tree, where each node of the tree is a Java object. Additional semantic information is attached to each node so that this data structure can be used for other purposes such as placement, partitioning, etc. Additional benefits of having an open and extensive VHDL compiler in Java are:

- Easiness of debugging VHDL descriptions
- Easiness of customizing the compiler to make more specific semantic analysis

Information related to the circuit graphical representation such as its size, placement of its components and connections routing can not be described in VHDL. New Java objects were created to hold this kind of information so they could be stored along with the VHDL description in the repository.

HardWWWired has two distinct ways of generating Java objects: the first one is through the VHDL/Java compiler as it has been explained in this section and the second one is through a method implemented in the Schematic Editor class (see Section 15.5.3) which generates the Java object tree adding graphical information.

One of the benefits of this mapping is that objects may have associated methods that allow different representation models of a system. The same object may have a graphical representation, textual representation, etc. These objects can be used in a variety of tools, such as: schematic editors, textual editors, browsers, etc. Additionally, changing parts of the system description and adding information to it becomes easier. Tools can handle objects in a more natural way. Finally creating generic components gets easier as the parameter passing becomes simpler. Each object corresponding to a generic component can have associated methods which performs automatic instantiation of the parameters.

15.4 SECURITY POLICIES

In an environment such as the Web some security policies must be taken for not allowing non-authorized access to private information. The HardWWWired architecture enables in a natural way employment of security policies. As it was pointed out earlier, the existence of filters attached to resources play an important role as controlling resource access.

200 VIRTUAL COMPONENTS DESIGN AND REUSE

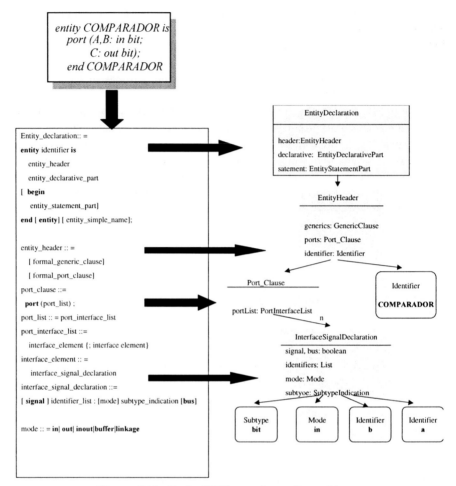

Figure 15.2 Entity BNF mapping to Java objects

When a user makes an HTTP request to some resource, before the request gets to the resource it passes through the resource filter. If the user has the rights to use that resource, the request will be granted, otherwise an error message will be sent back to the user informing the reason why the request could not be completed.

The prototype of HardWWWired uses a simple login-password scheme and an access rights policy similar to ordinary file systems such as the one found in UNIX systems.

15.5 DESIGNING A CPU WITH HARDWWWIRED: A CASE STUDY

There is a prototype of HardWWWired available at the URL: http://www.di.ufpe.br/~aams/codebase/examples/hardwwwired/. This prototype is entirely implemented in Java and can be executed in any browser that supports Java 1.1.x. It allows textual edition of VHDL descriptions with syntactic verification of it, creation of generic components, schematic edition of circuits with automatic generation of structural VHDL descriptions, and simulation and synthesis of structural and behavioral VHDL descriptions. Additionally, storage of VHDL components in the Resource Repository located at the server is also possible.

In order to better illustrates the reusability of components, an eight bit CPU, described in the next section, will be designed using HardWWWired in the next subsections.

15.5.1 Example Description

The example used to demonstrate how the prototype works consists of an eight bit CPU with two registers (A and B) and an internal memory whose instruction set is listed in Table 15.1. This example starts assuming that the control unit, memory and processing unit have already been designed using the platform. A simplified architecture of the CPU is illustrated in Figure 15.3. The input ports are clock (clk) and reset and the output ports (not detailed in Figure 15.3) are useful for simulation purposes and include: values of registers A and B, flags from the ALU, etc.

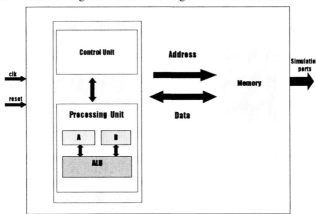

Figure 15.3 Simplified CPU architecture

Instruction	Opcode	Description	Length
NOP	00000	No operation	1
LDAA addr	00001	Load (addr) in Reg. A	2
LDAB addr	00010	Load (addr) in Reg. B	2
STAA addr	00011	Store Reg. A in addr.	2
STAB addr	00100	Store Reg. B in addr.	2
ADDA addr	00101	Reg. A <- Reg. A + (addr)	2
ADDB addr	00110	Reg. B <- Reg. B + (addr)	2
SUBA addr	00111	Reg. A <- Reg. A - (addr)	2
SUBB addr	01000	Reg. B <- Reg. B - (addr)	2
ADD	01001	Reg. A <- Reg. A + Reg. B	1
SUB	01010	Reg. A <- Reg. A - Reg. B	1
AND	01011	Reg. A <- Reg. A and Reg. B	1
NOTA	01100	Reg. A <- not Reg. A	1
NOTB	01101	Reg. B <- not Reg. B	1
CMP	01110	Compares Reg. A to Reg. B	1
JUMP addr	01111	Jump to address addr	2
JMPZ addr	10000	If zero, jump to addr	2
JMPC addr	10001	If carry, jump to addr	2
JMPN addr	10010	If negative, jump to addr	2
PUSH	10011	Push reg. A in stack	1
POP	10100	Pop stack top in reg.A	1
CALL addr	10101	Call subroutine	2
RET	10110	Return of subroutine	1

Table 15.1 CPU Instruction Set

15.5.2 Defining a New Component

The Schematic Editor is called by clicking on the Schematic Editor Button. Pressing down the New button, a dialog box appears asking the component's name and how many ports it has. After that another window appears and the user fills in the name, mode, type of each port of the component. This process is shown in Figure 15.4.

15.5.3 Schematic Edition with Automatic VHDL Code Generation

After defining the component it is time to build it using other components that are already in the Resource Repository. This is done by clicking the Insert button. A window appears listing all the Repository components. The components needed for

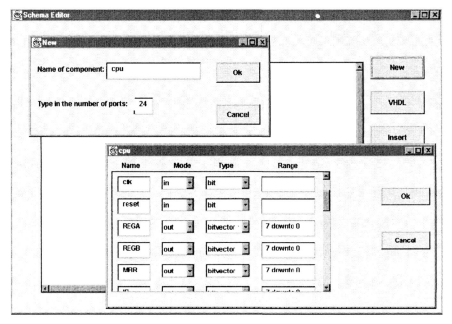

Figure 15.4 Creating a new component

the CPU are: a Control Unit (unidcontrolea), a Memory (memoria) and a Processing Unit (processador). After all components have been inserted, the next step is to connect them by dragging the mouse from one component port to the other. While all these steps are carried out automatic VHDL code generation occurs. To see the VHDL code generated, the user needs to click the VHDL button. This process is illustrated in Figure 15.5.

15.5.4 Simulation Environment: A New Front-End for the Alliance Design System

This prototype of HardWWWired includes the simulation and synthesis tools of the Alliance Design System [Grei93]. Although powerful, the simulation tool of Alliance has some problems:

- Test patterns - Input of test patterns is done by writing a text file obeying a syntax that can be understood by the simulator. This process is error prone.
- Simulation results representation - The simulation results are written in a text file. This makes harder the analysis process because generally there are a lot of inputs and outputs to be analyzed.

HardWWWired provides a new front-end for the Alliance Simulation tool. Input of test patterns is done by filling in a table and the results are presented in a wave

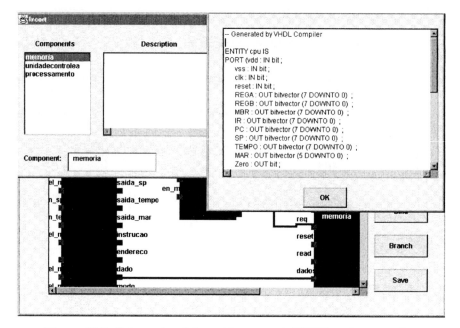

Figure 15.5 Schematic edition with automatic VHDL code generation

form. To simulate a VHDL description, the user has to click on the Simulation button and enter the test patterns. The simulation step is shown in Figure 15.6.

15.6 CONCLUSION AND FUTURE WORK

Issues concerning hardware components reusability and construction of hardware design environments are still complex. The Web is certainly a platform that offers many benefits for designing digital systems. Of course using the Web brings other issues to discussion such as security policies and performance, but it is still worthwhile. This work contributes to ease the digital systems design process by building a library of reusable components which can be shared among the universe of Web users. Another important advantage of it is to spread CAD tools technology to Web users, by creating an environment that brings most of CAD tools benefits at low cost. This environment also eases reusability by allowing definition of generic components and creation of new components from existing ones. This approach is quite different from other related works because it focuses on the main stages of the design process and uses VHDL as its base description language thus easing CAD tool interoperability. On the other hand, there are still issues not addressed in this work such as project management which plays an important role in design environments. However, important phases of the design process have been already

HARDWWWIRED: USING THE WEB AS REPOSITORY

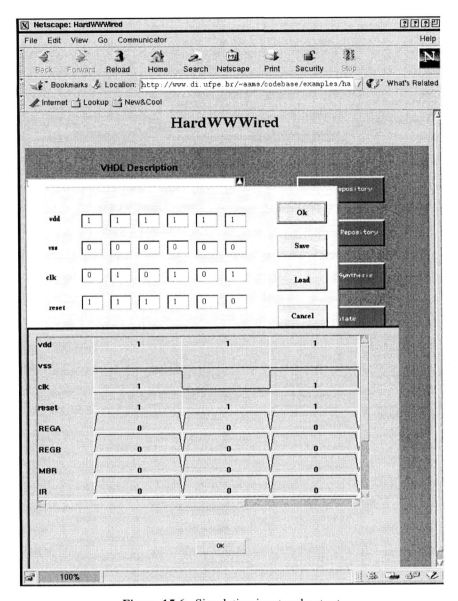

Figure 15.6 Simulation input and output

achieved. The prototype proves that the proposed HardWWWired architecture is feasible.

Future works include integrating more Web tools such as one for automatic schematic generation from a VHDL description, integrating more CAD tools in the Web server and also creating mechanisms for project management.

REFERENCES

[Akel91]   J. Akella and K. McMillan, "Synthesizing converters between finite state protocols", in *Proc. of ICCAD*, 1991.

[Ande88]   D.P. Anderson and P.V. Rangan, "High-Performance Interface Architectures for Cryptographic Hardware", in *Advances in Cryptology Proceedings - Eurocrypt'87*, Lecture Notes in Computer Science, Springer-Verlag, Berlin, pp. 301-309, 1988.

[Amme00]   J. Ammer, J. da Silva, Jr. et al., "Design Flow for Wireless Protocols", Berkeley Wireless Research Centre, January 2000.

[Arno96]   K. Arnold and J. Gosling, *The Java Programming Language*, Addison-Wesley, 1996.

[Ashe98]   P.J. Ashenden, P.A. Wilsey and D.E. Martin, "SUAVE: Object-Oriented and Genericity Extensions to VHDL for High-Level Modeling", in *Proc. of Forum on Design Languages (FDL)*, 1998.

[Bair96]   A. Baird-Smith, "Jigsaw: An object oriented server", W3 Consortium, MITLCS, USA. Jun, 1996. Available from `http://www.w3c.org`.

[Bare96]   L. Baresi, C. Bolchini and D. Sciuto, "Software Methodologies for VHDL Code Static Analysis based on Flow Graphs", in *Proc. of European Design Automation Conference (EuroDAC)*, 1996.

[Baum98]   L. Baum, L. Geyer, G. Molter, S. Rothkugel and P. Sturm, "Architecture-Centric Software Development Based on Extended Design Spaces", in *Proc. of ARES Second Int'l Workshop on Development and Evolution of Software Architectures for Product Families*, Las Palmas de Gran Canaria, Spain, February 1998

[Behn98]   B. Behnam, K. Babba and G. Saucier, "IP Taxonomy, IP Searching in a Catalog", in *Proc. of Design Automation and Test in Europe (DATE)*, Designer Track, pp. 147-151, 1998.

[Beie96]   Ch. Beierle, E. Börger, I. Durdanovic, U. Glässer and E. Riccobene, "Refining abstract machine specifications of the steam boiler control to well documented executable code", in J.-R. Abrial, E. Börger and H. Langmaack, editors, *Formal Methods for Industrial Applications: Specifying and Programming the Steam Boiler Control*, volume 1165 of Lecture Notes in Computer Science (State-of-the-Art Survey), pp. 52-78. Springer-Verlag, 1996.

[Berg99a]   R. Bergmann, S. Breen, M. Göker, M. Manago and S. Wess, "Developing industrial case-based reasoning applications: The INRECA methodology", in *Lecture Notes in Artificial Intelligence, 1612*, Springer, 1999.

[Berg99b] R. Bergmann and I. Vollrath, "Generalized Cases: Representation and Steps Towards Efficient Similarity Assessment", KI-99, in W. Burgard, A. B. Cremers and T. Christaller, editors, *Advances in Artificial Intelligence*, Proc. KI-99, Lecture Notes in Artificial Intelligence, Springer, 1999.

[Bern99] S. Bernardi et al., "Evaluation of VHDL-based design reuse through l-block analysis", in *Proc. of Forum on Design Languages (FDL)*,1999.

[Bigg93] T.J. Biggerstaff, B.G. Mitbander and D.Webster, "The concept assignment Problem in Program Understanding", in *Proc. of the Int. Conference on Software Engineering*, Baltimore, 1993.

[Bigg94] T.J. Biggerstaff, B.G. Mitbander and D.Webster, "Program understanding and the concept assignment Problem", in *Communications of the ACM*, 1994.

[Boek99] C. Böke and F.J. Rammig, "Design of Customizable Real-Time Communication Systems" (in German), in Journal *Informationstechnik und Technische Informatik (it+ti)*, 2, R. Oldenbourg Verlag, 1999.

[Boer95] E. Börger, U. Glässer and W. Müller, "Formal definition of an abstract VHDL'93 simulator by EA-machines", in C. Delgado Kloos and P.T. Breuer, editors, *Formal Semantics of VHDL*, volume 307 of The Kluwer International Series in Engineering and Computer Science, Kluwer Academic Publishers, 1995.

[Boer99] E. Börger, "High level system design and analysis using Abstract State Machines", in D. Hutter, W. Stephan, P. Traverso and M. Ullmann, editors, *Current Trends in Applied Formal Methods* (FM-Trends 98), Lecture Notes in Computer Science, Springer-Verlag, 1999.

[Boer98] E. Börger and J. Huggins, "Abstract State Machines 1988-1998: Commented ASM Bibliography", in *Bulletin of EATCS*, 64:105--127, February 1998.

[Bonn93] H. Bonnenberg and A. Curiger, "Vinci: Design and Test Report", *Technical Report No. 93/02*, Integrated Systems Laboratory, Swiss Federal Institute of Technology Zurich, 1993.

[Booc96] G. Booch, J. Rumbaugh and I. Jacobson, *The Unified Modeling Language for Object Oriented Development*, Version 1.0., 1996.

[Booc98] G. Booch, J. Rumbaugh and I. Jacobson, *The Unified Modeling Language Reference Manual*, Addison-Wesley, 1998.

[Bore97] J. Borel, "Technologies For Multimedia System On A Chip", in *IEEE International Solid-State Circuits Conference (ISSCC), Digest Of Technical Papers*, pp. 18-21, 1997.

[Bori88]   G. Boriello, "A new Interface Specification Methodology and Its Application to Transducer Synthesis", University of California at Berkeley, 1988.

[Bran99]   H.-J. Brand et al., "Redesign of an MPEG-2 HDTV video decoder considering reuse aspects", in R. Seepold and A. Kunzmann, editors, *Reuse Techniques for VLSI Design*, Kluwer Academic Publishers, 1999.

[Brin97]   O. Bringmann and W. Rosenstiel, "Cross-Level Hierarchical High-Level Synthesis", in *Proc. of Design Automation and Test in Europe (DATE)*, 1997.

[Brun99]   J.-Y. Brunel, E.A. de Kock, W.M. Kruijtzer, H.J.H.N. Kenter and W.J.M. Smits, "Communication Refinement in Video Systems on Chip", in *Proc. of the Intl. Workshop on Hardware/Software Codesign*, pp. 142-146, 1999.

[Calv93]   J.P.Calvez, *Embedded Real-Time Systems. A Specification and Design Methodology*, Ed. John Wiley 1993.

[Chak99]   S. Chakravarty and G. Martin, "A new embedded system design flow based on IP integration", in *Proc. of Design Automation and Test in Europe (DATE), User's Forum*, pp. 99-106, 1999.

[Chan99]   H. Chang, L. Cooke, M. Hunt, G. Martin, A. McNelly and L. Todd, *Surviving the SOC Revolution: A guide to platform-based design*, Kluwer Academic Publishers, 1999.

[Chao95]   L. Chaouat, C. Munk, A. Vachoux and D. Mlynek, "An Expert Assistant for Hardware Systems Specification", in *Proc. of Workshop on Libraries Component Modelling and Quality Assurance*, pp. 59-76, IRESTE - IHT, Nantes, France, April 26-27, 1995.

[Cher81]   L. A. Cherkasova and V. E. Kotov, "Structured nets", in J. Gruska and M. Chytil, editors, *Mathematical Foundations of Computer Science*, Volume 118 of Lecture Notes in Computer Science. Springer Verlag, 1981.

[Claa99]   T. A.C.M. Claasen, "High Speed: Not the Only Way to Exploit the Intrinsic Computational Power of Silicon", in *Proc. of International Solid State Circuits Conference*, pp. 22-25, 1999.

[Clar98]   E.M.Clarke, M.Fujita, S.P.Rajan, T. Reps, S.Shankar and T. Teitelbaum, "Program Slicing for Design Automation: An Automatic Technique for Speeding-up Hardware Design, Simulation and Verification", Computer Science Department, University of Wilsconsin, 1998.

[Cost99]   C. Costi and M. Miller, "A VHDL analysis environment for design reuse", in *Proc. of Forum on Design Languages (FDL)*, 1999.

[Cros79]   P.B. Crosby, *Quality is Free*, McGraw-Hill, New York, 1979.

[Cypr90]   Cypress Semiconductor, *SPARC RISC User's Guide*, ROSS Technology Inc., 1990.

[Dabr98a] J. Dabrowski and A. Pulka, "Discrete Approach to PWL Analog Modeling in VHDL Environment", in *Analog Integrated Circuits and Signals Processing*, Vol.16, No.2, pp. 3-11, Kluwer Academic Pub. 1998.

[Dabr98b] J. Dabrowski and A. Pulka, "Efficient Modeling of Analog and Mixed A/D Systems via Piece-wise Linear Technique", in *Proc. of Forum on Design Languages (FDL 98)*, pp. 295-304 Lausanne, 6-11 Sept.1998.

[Delf98]   P.Delforge, "IP Business Models", in *Proc. of the Intellectual Property System on Chip Conference (IP98)*, 1998.

[Demi86]   W.E. Deming, "Out of the Crisis", *MIT Centre for Advanced Engineering Studies*, Cambridge MA, 1986.

[Desi99]   DesignKit FHG8051, 1999.
           http://www.corepool.com/fhg8051.html

[Ditz98]   C. Ditze, "A Step Towards Operating System Synthesis", in *Proc. of the 5th Annual Australasian Conf. on Parallel and Real-Time Systems (PART)*, IFIP, IEEE, Adelaide, Aust., Sept. 1998.

[Eco96]   EDA Study supported by OMI (ECO 1996)

[Edaa97]   European Design and Automation Association (EDAA), *Roadmap System Design Technology*, November 1997.

[Eise99]   W. Eisenmann et al., "Hard IP reuse methodology for embedded cores", in R. Seepold and A. Kunzmann, editors, *Reuse Techniques for VLSI Design*, Kluwer Academic Publishers, 1999.

[Faul99]   N. Faulhaber and R. Seepold, "A Flexible Classification Model for Reuse of Virtual Components", in R. Seepold and A. Kunzmann, editors, *Reuse Technoques for VLSI Design*. Publishers, pp. 21-36, 1999

[Fiel]   R. Fielding et al., "RFC 2068: Hypertext Transfer Protocol - HTTP1.1", W3 Consortium, MITLCS, USA.
           Available from http://www.w3c.org.

[Fost98]   R. Foster and P. Young, "VLSI Velocity Rapid Silicon Prototyping: A Methodology to Simplify IP Integration and Verification", in *Proc. of IP 98 Europe*, pp. 117-151, 1998.

[Gast98]   Gastaldello, "800 Kgate in four Months: a Strategy for Reuse", in *Proc. of Intellectual Property (IP98)*, Munich 1998. Extension to the proceedings.

[Genr87]   H. J. Genrich, "Predicate/Transition Nets", in *Advances in Petri Nets Part I*. Volume 254 of Lecture Notes in Computer Science. Springer Verlag, 1987.

[Girc93]   E. Girczy and S. Carlson, "Increasing Design Quality and Engineering Productivity through Design Reuse", in *Proc. of Design Automation Conference (DAC)*, 1993.

[Glae99]   U. Glässer, R. Gotzhein and A. Prinz, "Towards a new formal SDL semantics based on abstract state machines", in G. v. Bochmann, R. Dssouli and Y. Lahav, editors, *9th SDL Forum Proceedings*, Elsevier Science B.V., 1999 (to appear).

[Grei93]   A. Greiner and F.Pêcheux, "A Complete Set of CAD Tools for Teaching VLSI Design", Laboratoire Mais/CAO-VLSI, Université Pierre et Marie Curie, Paris, France, 1993.

[Gutb91]   P. Gutberlet, H. Krämer and W. Rosenstiel, "CASCH - a Scheduling Algorithm forHigh Level -Synthesis", in *Proc. of the EDAC*, February 1991.

[Gutb94]   P. Gutberlet and W. Rosenstiel, "Timing Preserving Interface Transformations for the Synthesis of Behavioural VHDL", in *Proc. of EURO-DAC*, September 1994.

[Haas99a] J. Haase, T. Oberthür and M. Oberwestberg, "Design methodology for IP providers", in R. Seepold and A. Kunzmann, editors, *Reuse Techniques for VLSI Design*, Kluwer Academic Publishers, 1999.

[Haas99b] J. Haase, "Design methodology for IP providers", in *Proc. of Design Automation and Test in Europe (DATE)*, 1999.

[Halb93]   N. Halbwachs, *Synchronous Programming of Reactive Systems*, Kluwer Academic Publishers, 1993.

[Hans97]   C. Hansen, A. Kunzmann and W. Rosenstiel, "Verification by Simulation Comparison Using Interface Synthesis", in *Proc. of Design Automation and Test in Europe (DATE)*, 1997.

[Hare77]   D. Harel, "StateCharts: A visual formalism for complex systems", in *Science of Computer Programming* 8(3), pp. 231-274, 1977.

[Hash95]   M.M.Kamal Hashmi and Alistair C.Bruce, "Design and Use of a System-Level Specification and Verification Methodology", in *Proc. of the EDAC*, 1995.

[Hash97]   M.M.Kamal Hashmi, "Interface-based Design", Technical Note – distributed to the VSIA SLD DWG, 1997.

[Hash00]   M.M.Kamal Hashmi, *"VHDL+ Language Reference Manual"*, ICL Technical Report.

[Henz96]   T.A. Henzinger, "The theory of hybrid automata", in *Proc. of the 11th Annual Symposium on Logic in Computer Science*. IEEE Society Press, 1996.

[Heue97]   A. Heuer, *Objektorientierte Datenbanken, Konzepte, Modelle, Standards, und Systeme*, Addison-Wesley-Longman, 1997.

[Hodg97]   S.Hodgson and M.M.K.Hashmi, "SuperVISE - System Specification and Design Methodology", in *ICL Systems Journal* Vol. 12 Issue 2 November 1997.

[Hopc86]   J.E. Hopcroft and J.D. Ullman, *Introduction to Automata Theory, Languages, and Computation*, Addison-Wesley, 1986.

[Horw92]   S. Horwitz and T. Reps, "The Use of Program Dependence Graphs in Software Engineering", in *Proc. of the 14th International Conference on Software Engineering*, 1992.

[ICL97]   ICL Manchester, *VHDL+ - extensions to VHDL for interface specification*, 1997.

[ICL98]   ICL High Performance Systems, *VHDL+: Extensions to VHDL for System Specification*, 1998.

[IEEE93]   IEEE, *VHDL Language Reference Manual*, ANSI/IEEE Standard 1076-1993, June 1993.

[IEEE98]   IEEE, *P1076.6/D1.12 Draft Standard for VHDL Register Transfer Level Synthesis*, March 1998.

[IEEE99]   IEEE, *Std 1076.6 - Standard for VHDL Register Transfer Level Synthesis* (to be approved), IEEE Organization, 1999.

[ILog97]   I-Logix, *STATEMATE, User Reference Manual, Magnum 1.0*. I-Logix Inc, 1997.

[Indr97]   L. Indrusiak and R.Reis, "A WWW Approach for EDA Tool Integration", in *Proc. of X SBBCI*, pp 11-20, 1997

[ISI98]   ISI, "Design Automation Solutions". Integrated Systems Inc., 1998. http://www.isi.com/Products/DAS

[ISO87]   ISO8402, *Quality Vocabulary Part 1*, International Terms, 1987

[ISO91]   ISO/IEC International Standard 10116. *Information technology - Modes of operation for an n-bit block cipher algorithm*. 1991.

[ITUT92]   ITU-T, "ITU-T Specification and Description Language SDL", Recommendation Z.100 (SDL-91), ITU General Secretariat, Geneva, 1992. http://www.itu.ch/ITU-T

[Iwai96]   M. Iwaihara, M. Nomura, S. Ichinose and H. Yasuura, "Program Slicing on VHDL Descriptions and Its Applications", in *Proc. of the 3rd Asian Pacific Conference Hardware Description Languages*, Bangalore, 1996.

[Jebs93]   A.Jebson, C.Jones and H.Vosper, "CHISLE: An Engineer's tool for Hardware System Design", in *ICL Technical Journal* Vol. 8 No. 3 May 1993.

[Jerr97]   A. Jerraya, H. Ding, P. Kission and M. Rahmouni, "Behavioral Synthesis and Component", in R. Seepold and A. Kunzmann, editors, *Reuse with VHDL*, Kluwer Academic Publishers, 1997.

[Jone90]   C.B. Jones, *Systematic Software Development using VDM*, Prentice Hall, 1990.

[Jozw96]   L. Józwiak, "Modern Concepts of Quality and Their Relationship to Design reuse and Model Libraries", in *Hardware Component Modeling*, Kluwer Academic Publisher, 1996.

[Kaiz]   Kaizen. The key to Japan's competitive success

[Kate83]   M. Katevenis, *Reduced Instruction Set Computer Architecture for VLSI*, Ph. D. dissertation, Computer Science Div., Univ. of California, Berkeley, 1983.

[Keat98a] M. Keating and P. Bricaud, *Reuse Methodology Manual*, Kluwer Academic Publishers, 1998.

[Keat98b] M. Keating, "The necessary next step in Designer Productivity", in *Proc. of SNUG '98 Design Reuse*, 1998. `http://www.synopsys.com`

[Keat99]   M. Keating and P. Bricaud, *Reuse Methodology Manual: For Systems-On-A-Chip Design"*, (Second edition) Kluwer Academic Publishers, 1999.

[Kent99]   H.J.H.N. Kenter, C. Passerone, W.J.M. Smits, Y. Watanabe and A.L. Sangiovanni-Vincentelli, "Designing Digital Video Systems: Modeling and Scheduling", in *Proc. of International Workshop on Hardware/Software Codesign*, pp. 64-68, 1999.

[Kien99]   A.J.C. Kienhuis, *Design Space Exploration of Stream-Based Dataflow Architectures: Methods and Tools*, Ph. D. Thesis, TU Delft, January, 1999.

[Kiss95]   P. Kisson, H. Ding and A. Jerraya, "VHDL Based Design Methodology for Hierarchy and Component Re-Use", in *Proc. of EURO-DAC*, 1995.

[Kiss97]   P. Kission, A. Jerraya and I. Moussa, "Hardware Reuse", in *Proc. of the 2nd Workshop on Libraries Component Modelling and Quality Assurance with CHDL'97 and VHDL Forum*, pp.21-29, Toledo, Spain, April 20-25, 1997.

[Klein96] B. Kleinjohann, E. Kleinjohann and J. Tacken, "The SEA Language for System Engineering and Animation", in *Applications and Theory of*

	Petri Nets. Volume 1091 of Lecture Notes in Computer Science. Springer Verlag, 1996.
[Klein97]	B. Kleinjohann, J. Tacken and C. Tahedl, "Towards a Complete Design Method for Embedded Systems Using Predicate/Transition-Nets", in *Proc. of the XIII IFIP WG 10.5 Conference on Computer Hardware Description Languages and Their Applications (CHDL'97)*, Toledo, Spain, Chapman & Hall, 1997.
[Koeg98]	M. Kögst, P. Conradi, D. Garte and M. Wahl, "A Systematic Analysis of Reuse Strategies for Design of Electronic Circuits", in *Proc. of Design Automation and Test in Europe (DATE)*, 1998.
[Koeg99]	M. Koegst et. al., "IP retrieval by solving constraint satisfaction problems", in *Proc. of Forum on Design Languages (FDL)*, 1999.
[Leda98]	LEDA, *VHDL*Verilog System, Implementor's Guide*, v4.3.2, 1998.
[Leda00]	Web site of LEDA S.A., http://www.leda.fr. Now acquired by Synopsys http://www.synopsys.com.
[Lehm96a]	G. Lehmann, B. Wunder and K.D. Müller-Glaser, "A VHDL Reuse Workbench", in *Proc. of European Design Automation Conference*, 1996.
[Lehm96b]	G. Lehmann, B. Wunder and K.D. Muller-Glaser, "Basic Concepts for a HDL Reverse Engineering Tool-Set", in *Proc. of ICCAD-96*, 1996.
[Maff96]	O. Maffeïs, M. Morley and A. Poigné, "The Synchronous Approach to Designing Reactive Systems", in *Arbeitspapiere der GMD*, No. 973. GMD, St. Augustin, Germany, 1996.
[Mart98a]	G. Martin, A. McNelly and L. Todd, "The Integration Platform Approach to System-On-Chip Design", in *Proc. of IP 98 Europe*, pp. 101-116, 1998.
[Mart98b]	G. Martin and B. Salefski, "Methodology and Technology for Design of Communications and Multimedia Products via System-Level IP Integration", in *Proc. of Design Automation and Test in Europe (DATE), Designer Track*, pp. 11-18, 1998.
[Mart98c]	G. Martin, "System on a chip design for third generation wireless", in *Proc. of Electronic Product Design (EPD)*, pp. C22-C30, November, 1998.
[Mart98d]	G. Martin, "Moving IP to the System Level: What Will it Take?", in *Proc. of Embedded Systems Conference*, San Jose, Volume 4, pp. 243-256, 1998.
[Mart99a]	G. Martin, B. Jackson, Louis Pandula and Mika Nuotio, "Key components of a wireless terminal platform architecture", in *Mobile Handsets 99*, 1999.

[Mart99b]   G. Martin, "Productivity in VC reuse: Linking SOC platforms to abstract system design methodology", in *Proc. of Forum on Design Languages (FDL)*, 1999.

[Mast96]   M. Mastretti, M. Sturlesi and S. Tomasello, "Quality Measures and Analysis: A Way to improve VHDL Models", in *Hardware Component Modeling*, Kluwer Academic Publisher, 1996.

[Math97]   Y. Mathys and M. Morgan, "Design reuse: a methodology and implementation", in *Proc. of 2nd Workshop on Libraries Component Modelling and Quality Assurance with CHDL'97 and VHDL Forum*, pp.31-35, Toledo, Spain, April 20-25, 1997.

[Mazo95]   S. Mazor and P. Langstraat, *A Guide to VHDL,* 2nd. Edition, Kluwer Academic Publishers, 1995.

[Mcle97]   McLeod, "Tool Solution & Future Strategies", Presentation at the VSI Alliance User's meeting, 1997.

[Mede00]   Micro-Electronics Development for European Applications (MEDEA), *The MEDEA DESIGN AUTOMATION ROADMAP*, 2nd release, 2000. http://www.medea.org

[Meye96]   M. Meyer, *Finite Domain Constraints: Declarativity meets Efficiency - Theory meets Application*, DISKI 79, UB Stuttgart, Ph.D Thesis 1996/4232, 1996.

[Meye99]   V. Meyer zu Bexten and A. Stürmer, "Design reuse experiment for analog modules 'Dream'", in R. Seepold and A. Kunzmann, editors, *Reuse Techniques for VLSI Design*, Kluwer Academic Publishers, 1999.

[Mich92]   P. Michel, U. Lauther and P. Duzzy, *The Synthesis Approach to Digital System Design*, Kluwer Academic Publishers, 1992.

[Miln94]   R. Milner, "Computing is interaction", in B. Pehrson and I. Simon, editors, *Proc. of the IFIP 13th World Computer Congress 1994, Volume I: Technology and Foundations*, pp. 232-233. Elsevier Science Publishers B.V., 1994.

[Mode95]   Modelsim/Vsystem *User's Manual, VHDL Simulator for PC's Running Windows&Windows NT* ver.4.4, Model Technology Inc., USA, 1995.

[Mous99]   I. Moussa, M. D. Nava and A. A. Jerraya, "Analyzing the cost of design for reuse", in R. Seepold and A. Kunzmann, editors, *Reuse Techniques for VLSI Design*, Kluwer Academic Publishers, 1999.

[Nara95]   S. Narayan and D. Gajski, "Interfacing Incompatible Protocols using Interface Process Generation", in *Proc. of the 32nd Design Automation Conference*, 1995.

[Nist99]   National Institute of Standards and Technology. "Advanced Encryption Standard (AES) Development Effort", 1999.
http://csrc.nist.gov/encryption/aes/

[Ober96]   J.Oberg, A. Kumar and A. Hemani, "Grammar-based hardware synthesis of data communication protocols", in *Proc. of the 9th International Symposium on System Synthesis*, 1996.

[Obri97]   K. O'Brian and S. Maginot, "Reusability and Portability Improvements Through User-Defined VHDL Subsets", in *Proc. of 2nd Workshop on Libraries, Component, Modeling, and Quality Assurance*, Toledo, April 1997.

[Oehl98]   P. Oehler, I. Vollrath, P. Conradi and R. Bergmann, "Are you READEE for IPs?", in *Proc. of 2nd GI/ ITG/ GMM- Workshop "Reuse Techniques for VLSI Design"*, FZI-Bericht Karlsruhe. Forschungszentrum Informatik, 1998.

[Olco95a] S. Olcoz, L. Entrena and L. Berrojo, "VHDL Virtual Prototyping", in *Proc. of the 6th IEEE Int'l Workshop on Rapid System Prototyping*, pp. 161-167. Chapel Hill, NC, June 1995.

[Olco95b] S. Olcoz, L. Entrena and L. Berrojo, "An Effective System Development Environment based on VHDL Prototyping", in *Proc. of EuroDAC/EuroVHDL*, 1995.

[Olco98a] S. Olcoz, L. Ayuda, I. Izaguirre and O. Peñalba, "VHDL Teamwork, Organization and Workspace Management", in *Proc. of Design Automation and Test in Europe (DATE)*, 1998.

[Olco98b] S. Olcoz, A. Castellví and M. García, "Static Analysis Tools for Soft-Core Reviews and Audits", in *Proc. of Design Automation and Test in Europe (DATE)*, 1998.

[Olco98c] S. Olcoz, A. Castellví and M. García, "Improving VHDL Soft-Cores with Software-like Reviews and Audits Procedures", in *Proc. of VIUF'98*, Santa Clara, CA, pp. 143-146, 1998.

[Olco99]   S. Olcoz, A. Gutiérrez and D. Navarro, "A SPARC® mProcessor from VHDL to Silicon", in *Proc. of Design Automation and Test in Europe (DATE) (Users Forum)*, 1999.

[OMG99]   Object Management Group, *UML Profile for Scheduling, Performance and Time - Request for Proposal (ADTF RFP-9)*, OMG Document: ad/99-03-13, 1999.

[OVI95]   Open Verilog International, *Standard Delay Format Specification*, Version 3.0, May 1995.

… REFERENCES …

[Pass98]   R. Passerone, J.A. Rowson and A. Sangiovanni-Vincentelli, "Automatic Synthesis of Interfaces between Incompatible Protocols", in *Proc of the 35th Design Automation Conference*, 1998.

[Pete81]   J.L. Peterson, *Petri Net Theory and the Modelling of Systems*, Prentice Hall, 1981.

[Pibu94]   PI-BUS Specifications. Rev.0.3d, OMI324, OMI-Standards Project, 1994.

[Prei95a] V. Preis, R. Henfling, M. Schutz and S. März-Rössel, "A reuse scenario for the VHDL-Based Hardware Design Flow", in *Proc. of European Design Automation Conference*, pp. 464-469, 1995.

[Prei95b] V. Preis and S. März-Rössel, "Aspects of Modeling a Library of Complex and Highly Flexible Components in VHDL", in *Proc. of Workshop on Libraries Component Modelling and Quality Assurance*, pp.39-58, IRESTE - IHT, Nantes, France, April 26-27, 1995.

[PREN96]   PRENDA, "ASIC Design Methodology", *PRENDA Project*, February 1996.

[Prid95]   J. Pridmore. "The Standard Virtual Interface: An Interoperability Approach." *The RASSP Digest*, vol. 2, 4th Qtr. 1995.

[Pulk97]   A. Pulka and A. Pawlak, "Experiences with VITAL Code Generator Controlled by a Nonmonotonic Inference Engine", in *Proc. of 2nd Workshop on Libraries Component Modelling and Quality Assurance with CHDL'97 and VHDL Forum*, pp.309-320, Toledo, Spain, April 20-25, 1997.

[Quin99]   R. Quinnell, "Platforms for Integration", in *Silicon Strategies*, pp. 10-15, January, 1999.

[Raba99]   J. Rabaey, A. Sangiovanni-Vincentelli and R. Brodersen, "Communication/Component-Based Design and the PicoRadio Design Driver", *Gigascale Silicon Research Centre (GSRC) annual review*, December 9, 1999.

[Rade97]] M. Radetzki, W. Putzke-Röming and W. Nebel, "Objective VHDL: The Object-Oriented Approach to Hardware Reuse", in J.-Y. Roger, B. Stanford-Smith, P.T. Kidd, editors, *Advances in Information Technologies: The Business Challenge.* IOS Press, Amsterdam, 1998. Presented at EMMSEC'97, Florence, Italy, 1997.

[Rafi98]   R. Rafidinitrimo, P. Coeurdevey and G. Saucier, "An Object Oriented IP Managment System", in *Proc. of International Workshop on IP based synthesis and system design,* Grenoble, pp. 9-15, 1998.

[Rass99]   RASSP Education & Facilitation Program, Module 35.
           http://rassp.scra.org

[Rati97]   Rational, UML Summary, Version 1.1., 1997. http://www.rational.com/uml/resources/documentation/summary

[Reut97]   A. Reutter, B. Mößner, I. Kreuzer and W. Rosenstiel, "Wiederverwendung komplexer Komponenten für Synthese und Simulation unter Verwendung von VHDL", in *Proc. of the 8. E.I.S.- Workshop, GI/GMM/ITG Fachtagung*, pp. 105-114, 1997. In German.

[Reut99]   A. Reutter and W. Rosenstiel, "An Efficient Reuse System for Digital Circuit Design", in *Proc. of Design Automation and Test in Europe (DATE)*, 1999.

[Ries96]   T. Riesgo and J. Uceda, "A Fault Model for VHDL Descriptions at the Register Transfer Level", in *Proc. of EURO-DAC with EURO-VHDL'96*, Geneva, September 1996.

[Ries97]   T. Riesgo, Y. Torroja, C. Lopez and J. Uceda, "Estimation of the Quality of Design Validation Based on Error Models", in *Proc. of VUFE*, Toledo, April 1997.

[Rows97]   J.A. Rowson and A. Sangiovanni-Vincentelli, "Interface-Based Design", in *Proc. of the 31st Design Automation Conference*, 1997.

[Rust98]   C. Rust, J. Stroop and J. Tacken, "The Design of Embedded Real-Time Systems using the SEA Environment", in *Proc. of the 5th Annual Australasian Conference on Parallel And Real-Time Systems (PART '98)*, Adelaide, Australia, 1998.

[Sang96]   A.Sangiovanni-Vincentelli, P.C.McGeer and A.Saldanha, "Verification of Electronic Systems", in *Proc. of Design Automation Conference (DAC)*, 1996.

[Sant00]   M. Santarini, "Cadence rolls system-level design to fore", in *Electronic Engineering Times*, pp. 1,22,24,130, January 10, 2000.

[Schl99]   U. Schlichtmann and B. Wurth, "An integrated approach towards a corporate design reuse strategy", in R. Seepold and A. Kunzmann, editors, *Reuse Techniques for VLSI Design*, Kluwer Academic Publishers, 1999.

[Schn94]   B. Schneier, *Applied Cryptography, Protocols, Algorithms, and Source Code in C*, John Wiley & Sons, 1994.

[Schu95]   G. Schumacher and W. Nebel, "Inheritance Concept for Signals in Object-Oriented Extensions to VHDL", in *Proc. of European Design Automation Conference*, 1995.

[Seep98]   R. Seepold and N. Faulhaber, "An Efficient Similarity Metric for IP Reuse in RMS", in *Proc. of 2nd GI/ ITG/ GMM- Workshop "Reuse Techniques for VLSI Design"*, FZI-Bericht Karlsruhe. Forschungszentrum Informatik, 1998.

[Seep99a]	R. Seepold, "Reuse of IP and Virtual Components", in *Proc. of Design Automation and Test in Europe (DATE)*, 1999.
[Seep99b]	R. Seepold and A. Kunzmann, *Reuse Techniques for VLSI Design*, Kluwer Academic Publishers, March 1999. ISBN 0-7923-8476-8.
[Seep00]	R. Seepold, N. Martínez Madrid and W. Rosenstiel, "Reuse of Virtual Components in System-on-Chip Environments", in *Tutorials of Design Automation and Test in Europe (DATE)*, March 2000.
[Seli99]	B. Selic, "Turning clockwise: using UML in the real-time domain", in *Communications of the ACM*, pp. 46-54, October 1999.
[SIA99]	Semiconductor Industry Association, *International Technology Roadmap for Semiconductors: 1999 edition*. Austin, TX:International SEMATECH, 1999.
[Sica99a]	SICAN DesignObjects™ Catalogue and Handouts: http://www.sican.de/ (look for Products, DesignObjects), 1999.
[Sica99b]	SICAN DesignObjects™ Deliverables and Integration Process: http://www.sican.de/ (look for Products, DesignObjects), 1999.
[Sieg98]	R. Siegmund, H. v. Sychowski, J. Lancaster and D. Mueller, "Specification and verification of complex digital systems using VHDL+", in *Proc. of IP'98*, 1998.
[Sieg99]	R. Siegmund and D. Mueller, "An approach to specification and synthesis of adaptive interfaces of reusable hardware modules", in *Proc. of Forum on Design Languages (FDL)*, 1999.
[Sina94]	P. Sinander, "VHDL Modeling Guidelines", Issue 1, *European Space Agency Internal Document*, September 1994.
[Slam99]	D. Slama, J. Garbis and P. Russell, *Enterprise CORBA*, Prentice-Hall, Upper Saddle River, 1999.
[Smit98]	J. Smith and G. de Micheli, "Automated Composition of Hardware Components", in *Proc. of the 35th Design Automation Conference*, 1998.
[Spar91]]	SPARC International Inc., *SPARC Architecture Version 8*, Prentice Hall, 1991.
[Spar92]	SPARC International Inc., *SPARC Compliance Definition SCD*, SI Inc., 1992.
[Spar94]	SPARC International Inc., *SPARC-V8 Embedded (V8E) Architecture Specification*, SI Inc., 1994.
[Sunt98]	Suntest, *Java Compiler: Compiler The Java Parser Generator*, Sun Microsystems, Inc. CA USA, 1998. Available from http://www.sun.com

[Swam95]   S. Swamy, A. Molin and B. Govnot, "OO-VHDL: Object-oriented extensions to VHDL", in *IEEE Computer*, vol. 28, no.10, October 1995.

[Tamm98]   S. Tamme, "IP Business Models - what is the real value of Intellectual Property?", in *Proc. of the Intellectual Property System on Chip Conference (IP98)*, 1998.

[Tanu95]   Y. Tanurhan, S. Schmerler and K.D. Müller-Glaser, "A Backplane Approach for Cosimulation in High-Level System Specification Environments", in *Proc. of EURO-DAC*, 1998.

[Tich91]   F.W. Tichy, "RCS - A System of Version Control, Department of Computer Sciences", Purdue University, 1991.

[TOMI96]   TOMI, "Report on Modelling, Design and Testability Quality Metrics", *TOMI Internal Report*, ESPRIT Project 20724, 1996.

[Torr97]   Y. Torroja, T. Riesgo, E. de la Torre and J. Uceda, "Design Reusability: Generic and Configurable Designs", in *Proc. of VUFE*, Toledo, April 1997.

[Torr99]   Y. Torroja, "ARDID: A tool for the quality analysis of VHDL based designs", in *Proc. of Forum on Design Languages (FDL)*, 1999.

[Tsan93]   E. Tsang, *Foundations of Constraint Satisfaction*, Academic Press, Harcourt Brace & Company, Publishers, 1993.

[Usse93]   Rudolf Usselmann, *MicroSPARC Instruction Set Architecture Test Suite Manual*, SPARC International, 1993.

[VCX00]   Virtual Component Exchange, 2000. http://www.vcx.org

[VITA95]   VITAL'95 ASIC Modeling Specification, *IEEE 1076.4, VHDL Initiative Towards ASIC Libraries*, The Institute of Electrical and Electronics Engineers, Inc., New York, October 1995.

[VSIA96]   Virtual Socket Interface Alliance, *Architecture document*, 1996. http://www.vsi.org

[VSIA98]   Virtual Socket Interface Alliance, *VSI System Level Design Model Taxonomy*, Version 1.0, October 1998.

[VSIA00a]   VSI Alliance, *Virtual Component Interface Standard (OCB 2 1.0)*, 2000.

[VSIA00b]   Virtual Socket Interface Alliance, http://www.vsi.org

[VSIA00c]   Virtual Socket Interface Alliance, *System Level Interface Behavioural Documentation Standard*, Interface subgroup of VSIA System Level Design DWG, currently in development - contact C. Lennard (chair).

[Wagn94]   R.F. Wagner, "Ambientes de Projeto de Sistemas Eletrônicos", Instituto de Informática UFRGS, 1994.

[Wegn97] P. Wegner, "Interaction is more powerful than algorithms", in *Communications of the ACM* 40(5): 81-92, May 1997.

[Weld] U. C. Berkeley, *WELD Project.*
http://www.cad.eecs.berkeley.edu/weld

[Wilk99] D. Wilkes and M.M.Kamal Hashmi, "Application of High Level Interface-based Design to Telecommunications System Hardware", in *Proc. of Design Automation Conference (DAC)*, 1999.

[Wolf94] W. H. Wolf, "Hardware-software co-design of embedded systems", in *Proc. of the IEEE*, 7(82):967--989, 1994.

[XML98] XML, *Extensible Markup Language (XML) 1.0,* W3 Consortium, 1998.
http://www.w3.org/TR/1998/REC-xml-19980210.html

WICHTIG:
[Olco98] ersetzt duch [Olco98a]
[Prei98] ersetzt durch [Prei98a]

INDEX

A

A/D 172
abstraction 172
API 113
artificial intelligence 106, 171
ASI 185
ASM 47, 53
ATE 188
automated component generator 171

B

behavioral synthesis 131
business model 22

C

C 154
C++ 154
CADDY-II 136
CBR 106
CE 18
CFG 87
CFSM 36
classification 24, 27
computational efficiency see CE
concept assignment problem 85
concurrent engineering 23
CORBA 47
core
    programmable 35
    soft 120
cryptographic function 119

D

DASC 145
dataflow 85
DBM 112
DCOM 48
design efficiency 23
design flow 5, 22
Design For Manufacturability 35

Design For Test 35
design methodology 22
design reuse 22
design-ins 1
DFG 87

E

EDA 95
efficiency
    synthesis 66
ER-diagram 113
executable model 146

F

formal verification 5
FPDG 85
    control dependency edges 87
    data dependence edges 87
    predicate node 87
    region node 87
    statement node 87
FPU 185
frame component 134
functional analysis 86, 92
functional task model 38

G

garbage administration 28
GEM 171
GIOP 49

H

hardware/software co-design 14
HDL 82, 172
high-level synthesis 131

I

ICE 18
IDL 49

IEEE
    IEEE1076 98
    IEEE1076.6 74, 88
IEEE standard (PAR. 1754) 185
IIOP 49
Integer Unit see IU
integration 172
integration platform 34
Intellectual property see IP
interface
    customization 159
    hardware-hardware 153
    hardware-software 153
    software-software 153
    specification 162
    synthesis 166
    taxonomy 155
Intrinsic computational efficiency see ICE
IP 14, 106, 131, 146, 183
    3rd party 29
    archiving 107
    classes 107
    contract 29
    life-cycle model 28
    pre-qualified 34
    protection 28, 31
    qualification 26
    qualified 25
    repository 162
    retrieval 106
    reuse 22
    soft 159
    software 47
    unqualified 25
IU 185

J

JESSI 32

L

lambda-block 96
legacy design 82

M

MA 171, 172, 178
MEDEA 13
mixed-signal design 22
MMU 185
MSC 154

N

NVMG 172

O

object-oriented techniques 24
OMG 46, 48

P

PDG 87
prototyping 22
PWL 172

Q

quality
    complexity 66
    documentation 96
    maintainability 66
    measurement 28
    portability 66
    product 65
    readability 66
    reusability 66
    test bench 71

R

rapid prototyping 24
RCOS 48
RCS 67
REM 171
reuse
    as is 81
    hierarchical 180
    horizontal 24
    inter-company 26

INDEX   225

intra-company 26
IP-based 81
library 119
vertical 24
RMI 50, 112
RMS 23
royalty 30
RT 131
RTOS 35, 48

S

scalable bus 127
SCD 185, 188
SDF 37
SDL 154
secure communication 125
selection 172
SGML 57
signal analysis 86, 89
similarity detection 24
similarity function 106
simulation performance 66
slicing 82
SLIF 155, 157
SOC 14, 33, 145, 159, 183
soft component 82
specialization 172
specification 150
standard
   wireless
      CDMA 35
      GSM 35
system level design 34
System-on-Chip see SOC

T

testbench 157, 178
time to market 23, 184
TLI 162

U

UML 28, 46, 51

V

VC 1, 119, 145
   black-box 148
   cost 146
   exchange 154
   firm 146
   hard 146
   portfolio 35
   provider 154
   soft 146
   user 154
VCI 154
VCM 175
verification 24
Verilog 154
VHDL 5, 120
   entity 98
   legacy code 94
   VHDL+ 159
VHDL+ 154
Virtual Component Interface see VCI
Virtual Component see VC
virtual interface 121
virtual prototyping 183
VITAL 171
VOP 43
VSI 24, 30, 35
   compliant 35
   SLD 145
   VCT 30
   Virtual Component Interface 40
VSIA see VSI

W

workflow 5, 55

X

XMI 51
XML 47

Y

YAPI 39
Y-chart 36

GLOSSARY

A/D	Analog/Digital
AI	Articial Intelligence
API	Application Programming Interface
ASI	Address space identifier
ASM	Abstract State Machine
ATE	Automatic test equipment
CBR	Case-based reasoning
CE	computational efficiency
CFG	Control flow graph
CFSM	Co-design finite state machines
CORBA	Common Object Request Broker Architecture
CVC	Cryptographic virtual component
DASC	Design Automation Standards Committee
DBM	Data-Base-Management
DCOM	Distributed Common Object Model
Design-ins	Integration of IP module in ASIC design
DesignObjects	Sican's Trademark for Virtual Component
DFG	Data flow graph
DTD	Document Type Definition
EDA	Electronic Design Automation
ER	Entity-Relationship
FPDG	Finer Program Dependence Graph
FPU	Floating Point Unit
GEM	Generation Entity of the Model
GIOP	General Inter-ORB Protocol
HDL	Hardware Description Language
ICE	Intrinsic computational efficiency
IDL	Interface Definition Language
IIOP	Internet Inter-ORB Protocol
IU	Integer Unit
MA	Modelling assistant

MEDEA	Micro-Electronics Development for European Applications
MMU	Memory Management Unit
MSC	Message Sequence Charts
NVMG	Nonmonotonic VITAL Models Generator
OMG	Object Management Group
OMT	Object Management Technique
PDG	Program dependence graph
PWL	Piece-wise linear
RCOS	Real Time Communication System
RCS	Revision Control System
REM	Reuse Engine Module
RMI	Remote Method Invocation
RMS	Reuse management system
RT	Register Transfer
RTOS	Real-Time Operating System
SCD	SPARC Compliance Definition
SDF	Scheduled dataflow models
SGML	Standard Generalized Markup Language
SLD	System level design
SLIF	System-Level Interface Behavioral Documentation Standard
SOC	System-on-chip design
SQL	Structured Query Language
TLI	Transaction Level Interface
TTM	Time-to-market
UML	Unified Modelling Language
VC	Virtual Component
VCM	Virtual Component Model
VHDL	VHSIC - Very High Speed Integrated Circuit - Hardware Description Language
VITAL	VHDL Initiative Towards ASIC Libraries
VOP	Voice over Packet
VSI	Virtual Socket Interface Alliance

XMI	XML Metadata Interchange
XML	eXtensible Markup Language
XSL	eXtended Style Language
YAPI	Y-chart Application Programmers Interface